Molecular Biology and Biochemistry

Molecular Biology and Biochemistry

Dr. Nitin Suri

Oxford Book Company
Jaipur ▪ New Delhi

Published By

Oxford Book Company

Regd. Off. "Bony Residency", Gate No. 2
Opp. Tilak Public School, Vishweswariya Nagar,
Gopalpura Road, Jaipur - 302018, India
Ph: 0141 - 2761280, Telefax: 2761381
email - oxfordbookjpr@gmail.com
website - oxfordbookcompany.com

Delhi Office: 102, 1st Floor, Satyam House,
4327/3, Ansari Road, Darya Ganj,
New Delhi - 110002, India
Ph: 011-45652440

Distributed by
KEYA INTERNATIONAL PRESS LLP
E-mail: keyainternationalpressllp@gmail.com

Published in India, 2025

ISBN : 9789380179582

Typeset by Shivangi Computers, Jaipur

Printed and bound in India by Replika Press Pvt. Ltd.

Preface

Molecular biology is the study of biology at a molecular level. The field overlaps with other areas of biology and chemistry, particularly genetics and biochemistry. Molecular biology chiefly concerns itself with understanding the interactions between the various systems of a cell, including the interactions between DNA, RNA and protein biosynthesis as well as learning how these interactions are regulated. Researchers in molecular biology use specific techniques native to molecular biology, but increasingly combine these with techniques and ideas from genetics and biochemistry. There is not a defined line between these disciplines. Much of the work in molecular biology is quantitative, and recently much work has been done at the interface of molecular biology and computer science in bioinformatics and computational biology. The study of gene structure and function, molecular genetics, has been amongst the most prominent sub-field of molecular biology.

Increasingly many other fields of biology focus on molecules, either directly studying their interactions in their own right such as in cell biology and developmental biology, or indirectly, where the techniques of molecular biology are used to infer historical attributes of populations or species, as in fields in evolutionary biology such as population genetics and phylogenetics. There is also a long tradition of studying biomolecules "from the ground up" in biophysics.

This book offers a complete introduction to biochemistry and molecular biology. It covers both current and traditional areas in biochemistry, with an emphasis on the biological aspects of biochemistry. Adopting a readable, friendly style, complete with clear, full-color illustrations throughout, this book provides the essential information necessary for a complete understanding of biochemistry and molelcular biology. This is a very impressive well-organised, beautifully-illustrated introductory textbook. It should prove popular among a very broad variety of students for whom understanding of principles is more important than memorisation of large amounts of poorly comprehended factual information. There are many such students around the world who will benefit from this masterly book.

Dr. Nitin Suri

Contents

1

Introduction

Molecular biology is the study of biology at a molecular level. The field overlaps with other areas of biology and chemistry, particularly genetics and biochemistry. Molecular biology chiefly concerns itself with understanding the interactions between the various systems of a cell, including the interactions between DNA, RNA and protein biosynthesis as well as learning how these interactions are regulated. Researchers in molecular biology use specific techniques native to molecular biology, but increasingly combine these with techniques and ideas from genetics and biochemistry. There is not a defined line between these disciplines. The following figure is a schematic that depicts one possible view of the relationship between the fields:

Biochemistry is the study of the chemical substances and vital processes occurring in living organisms. Biochemists focus heavily on the role, function, and structure of biomolecules. The study of the chemistry behind biological processes and the synthesis of biologically active molecules are examples of biochemistry.

Genetics is the study of the effect of genetic differences on organisms. Often this can be inferred by the absence of a normal component (e.g. one gene). The study of "mutants" – organisms which lack one or more functional components with respect to the so-called "wild type" or normal phenotype. Genetic interactions (epistasis) can often confound simple interpretations of such "knock-out" studies.

Molecular biology is the study of molecular underpinnings of the process of replication, transcription and translation of the genetic material. The central dogma of molecular biology where genetic material is transcribed into RNA and then translated into protein, despite being an oversimplified picture of molecular biology, still provides a good starting point for understanding the field. This picture, however, is undergoing revision in light of emerging novel roles for RNA.

Much of the work in molecular biology is quantitative, and recently much work has been done at the interface of molecular biology and computer science in bioinformatics and computational biology. As of the early 2000s, the study of gene structure and function, molecular genetics, has been amongst the most prominent sub-field of molecular biology.Increasingly many other fields of biology focus on molecules, either directly studying their interactions in their own right such as in cell biology and developmental biology, or indirectly, where the techniques of molecular biology are used to infer historical attributes of populations or species, as in fields in evolutionary biology such as population genetics and phylogenetics. There is also a long tradition of studying biomolecules "from the ground up" in biophysics.

TECHNIQUES OF MOLECULAR BIOLOGY

Since the late 1950s and early 1960s, molecular biologists have learned to characterize, isolate, and manipulate the molecular components of cells and organisms. These components include DNA, the repository of genetic information; RNA, a close relative of DNA whose functions range from serving as a temporary working copy of DNA to actual structural and enzymatic functions as well as a functional and structural part of the translational apparatus; and proteins, the major structural and enzymatic type of molecule in cells.

Expression Cloning

One of the most basic techniques of molecular biology to study protein function is expression cloning. In this technique, DNA coding for a protein of interest is cloned (using PCR and/or restriction enzymes) into a plasmid (known as an expression vector). This plasmid may have special promoter elements to drive production of the protein of interest, and may also have antibiotic resistance markers to help follow the plasmid.This plasmid can be inserted into either bacterial or animal cells. Introducing DNA into bacterial cells can be done by transformation (via uptake of naked DNA), conjugation (via cell-cell contact) or by transduction (via viral vector). Introducing DNA into eukaryotic cells, such as animal cells, by physical or chemical means is called transfection. Several different transfection techniques are available, such as calcium phosphate transfection,electroporation, microinjection and liposome transfection. DNA can also be introduced into eukaryotic cells using viruses or bacteria as carriers, the latter is sometimes called bactofection and in particular uses Agrobacterium tumefaciens. The plasmid may be integrated into the genome, resulting in a stable transfection, or may remain independent of the genome, called transient transfection.

In either case, DNA coding for a protein of interest is now inside a cell, and the protein can now be expressed. A variety of systems, such as inducible promoters and specific cell-signaling factors, are available to help express the protein of interest at high levels. Large quantities of a protein can then be extracted from the bacterial or eukaryotic cell. The protein can be tested for enzymatic activity under a variety of situations, the protein may be crystallized so its tertiary structure can be studied, or, in the pharmaceutical industry, the activity of new drugs against the protein can be studied.

Polymerase Chain Reaction (PCR)

The polymerase chain reaction is an extremely versatile technique for copying DNA. In brief, PCR allows a single DNA sequence to be copied (millions of times), or altered in predetermined ways. For example, PCR can be used to introduce restriction enzyme sites, or to mutate (change) particular bases of DNA, the latter is a method referred to as "Quick change". PCR can also be used to determine whether a particular DNA fragment is found in a cDNA library. PCR has many variations, like reverse transcription PCR (RT-PCR) for amplification of RNA, and, more recently, real-time PCR (QPCR) which allow for quantitative measurement of DNA or RNA molecules.

Gel Electrophoresis

Gel electrophoresis is one of the principal tools of molecular biology. The basic principle is that DNA, RNA, and proteins can all be separated by means of an electric field. In agarose gel electrophoresis, DNA and RNA can be separated on the basis of size by running the DNA through an agarose gel. Proteins can be separated on the basis of size by using an SDS-PAGE gel, or on the basis of size and their electric charge by using what is known as a 2D gel electrophoresis.

Macromolecule Blotting and Probing

The terms *northern*, *western* and *eastern* blotting are derived from what initially was a molecular biology joke that played on the term *Southern blotting*, after the technique described by Edwin Southern for the hybridisation of blotted DNA. Patricia Thomas, developer of the RNA blot which then became known as the *northern blot* actually didn't use the term. Further combinations of these techniques produced such terms as *southwesterns* (protein-DNA hybridizations), *northwesterns* (to detect protein-RNA interactions) and *farwesterns* (protein-protein interactions), all of which are presently found in the literature.

Southern Blotting

Named after its inventor, biologist Edwin Southern, the Southern blot is a method for probing for the presence of a specific DNA sequence within a DNA sample. DNA samples before or after restriction enzyme digestion are separated by gel electrophoresis and then transferred to a membrane by blotting via capillary action. The membrane is then exposed to a labeled DNA probe that has a complement base sequence to the sequence on the DNA of interest. Most original protocols used radioactive labels, however non-radioactive alternatives are now available. Southern blotting is less commonly used in laboratory science due to the capacity of other techniques, such as PCR, to detect specific DNA sequences from DNA samples. These blots are still used for some applications, however, such as measuring transgene copy number in transgenic mice, or in the engineering of gene knockout embryonic stem cell lines.

Northern Blotting

The northern blot is used to study the expression patterns of a specific type of RNA molecule as relative comparison among a set of different samples of RNA. It is essentially a combination of denaturing RNA gel electrophoresis, and a blot. In this process RNA is separated based on size and is then transferred to a membrane that is then probed with a labeled complement of a sequence of interest. The results may be visualized through a variety of ways depending on the label used; however, most result in the revelation of bands representing the sizes of the RNA detected in sample. Thè intensity of these bands is related to the amount of the target RNA in the samples analyzed. The procedure is commonly used to study when and how much gene expression is occurring by measuring how much of that RNA is present in different samples. It is one of the most basic tools for determining at what time, and under what conditions, certain genes are expressed in living tissues.

Western Blotting

Antibodies to most proteins can be created by injecting small amounts of the protein into an animal such as a mouse, rabbit, sheep, or donkey (polyclonal antibodies)or produced in cell culture (monoclonal antibodies). These antibodies can be used for a variety of analytical and preparative techniques. In western blotting, proteins are first separated by size, in a thin gel sandwiched between two glass plates in a technique known as SDS-PAGE (sodium dodecyl sulfate polyacrylamide gel electrophoresis). The proteins in the gel are then transferred to a PVDF, nitrocellulose, nylon or other support membrane. This membrane can then be probed with solutions of antibodies. Antibodies that specifically bind to the protein of interest can then be visualized by a variety of techniques, including colored products, chemiluminescence, or autoradiography. Often, the antibodies are labeled with an enzymes. When a chemiluminescent substrate is exposed to the enzyme it allows detection. Using western blotting techniques allows not only detection but also quantitative analysis.

Analogous methods to western blotting can be used to directly stain specific proteins in live cells or tissue sections. However, these *immunostaining* methods, such as FISH, are used more often in cell biology research.

Eastern Blotting

Eastern blotting technique is to detect post-translational modification of proteins. Proteins blotted on to the PVDF or nitrocellulose membrane are probed for modifications using specific substrates.

Arrays

A DNA array is a collection of spots attached to a solid support such as a microscope slide where each spot contains one or more single-stranded DNA oligonucleotide fragment. Arrays make it possible to put down a large quantity of very small (100 micrometre diameter) spots on a single slide. Each spot has a DNA fragment molecule that is complementary to a single DNA sequence (similar to Southern blotting). A variation of this technique allows the gene expression of an organism at a particular stage in development to be qualified (expression profiling). In this technique the RNA in a tissue is isolated and converted to labeled cDNA. This cDNA is then hybridized to the fragments on the array and visualization of the hybridization can be done. Since multiple arrays can be made with the exact same position of fragments they are particularly useful for comparing the gene expression of two different tissues, such as a healthy and cancerous tissue. Also, one can measure what genes are expressed and how that expression changes with time or with other factors.

For instance, the common baker's yeast, *Saccharomyces cerevisiae*, contains about 7000 genes; with a microarray, one can measure qualitatively how each gene is expressed, and how that expression changes, for example, with a change in temperature. There are many different ways to fabricate microarrays; the most common are silicon chips, microscope slides with spots of ~ 100 micrometre diameter, custom arrays, and arrays with larger spots on porous membranes (macroarrays). There can be anywhere from 100 spots to more than 10,000 on a given array.

Allele Specific Oligonucleotide

Allele specific oligonucleotide (ASO) is a technique that allows detection of single base mutations without the need for PCR or gel electrophoresis. Short (20-25 nucleotides in length), labeled probes are exposed to the non-fragmented target DNA. Hybridization occurs with high specificity due to the short length of the probes and even a single base change will hinder hybridization. The target DNA is then washed and the labeled probes that didn't hybridize are removed. The target DNA is then analyzed for the presence of the probe via radioactivity or fluorescence. In this experiment, as in most molecular biology techniques, a control must be used to ensure successful experimentation. The Illumina Methylation Assay is an example of a method that takes advantage of the ASO technique to measure one base pair differences in sequence.

Antiquated Technologies

In molecular biology, procedures and technologies are continually being developed and older technologies abandoned. For example, before the advent of DNA gel electrophoresis (agarose or polyacrylamide), the size of DNA molecules was typically determined by rate sedimentation in sucrose gradients, a slow and labor-intensive technique requiring expensive instrumentation; prior to sucrose gradients, viscometry was used. Aside from their historical interest, it is often worth knowing about older technology, as it is occasionally useful to solve another new problem for which the newer technique is inappropriate.

DEVELOPMENT OF MOLECULAR BIOLOGY

The history of molecular biology begins in the 1930s with the convergence of various, previously distinct biological disciplines: biochemistry, genetics, microbiology, and virology. With the hope of understanding life at its most fundamental level, numerous physicists and chemists also took an interest in what would become molecular biology. In its modern sense, molecular biology attempts to explain the phenomena of life starting from the macromolecular properties that generate them. Two categories of macromolecules in particular are the focus of the molecular biologist:

1) nucleic acids, among which the most famous is deoxyribonucleic acid (or DNA), the constituent of genes, and

2) proteins, which are the active agents of living organisms.

One definition of the scope of molecular biology therefore is to characterize the structure, function and relationships between these two types of macromolecules. This relatively limited definition will suffice to allow us to establish a date for the so-called "molecular revolution", or at least to establish a chronology of its most fundamental developments.

In its earliest manifestations, molecular biology—the name was coined by Warren Weaver of the Rockefeller Foundation in 1938—was an ideal of physical and chemical explanations of life, rather than a coherent discipline. Following the advent of the Mendelian-chromosome theory of heredity in the 1910s and the maturation of atomic theory and quantum mechanics in the 1920s, such explanations seemed within reach. Weaver and others encouraged (and funded) research at the intersection of biology, chemistry and physics, while prominent physicists such as Niels Bohr and Erwin Schrödinger turned their attention to biological speculation. However, in the 1930s and 1940s it was by no means clear which—if any—cross-disciplinary research would bear fruit; work in colloid chemistry, biophysics and radiation biology, crystallography, and other emerging fields all seemed promising. In 1940, George Beadle and Edward Tatum demonstrated the existence of a precise relationship between genes and proteins. In the course of their experiments connecting genetics with biochemistry, they switched from the genetics mainstay *Drosophila* to a more appropriate model organism, the fungus *Neurospora*; the construction and exploitation of new model organisms would become a recurring theme in the development of molecular biology. In 1944, Oswald Avery, working at the Rockefeller Institute of New York, demonstrated that genes are made up of DNA. In 1952, Alfred Hershey and Martha Chase confirmed that the genetic material of the bacteriophage, the virus which infects bacteria, is made up of DNA. In 1953, James Watson and Francis Crick discovered the double helical structure of the DNA molecule.

In 1961, Francois Jacob and Jacques Monod hypothesized the existence of an intermediary between DNA and its protein products, which they called messenger RNA. Between 1961 and 1965, the relationship between the information contained in DNA and the structure of proteins was determined: there is a code, the genetic code, which creates a correspondence between the succession of nucleotides in the DNA sequence and a series of amino acids in proteins. At the beginning of the 1960s, Monod and Jacob also demonstrated how certain specific proteins, called regulative proteins, latch onto DNA at the edges of the genes and control the transcription of these genes into messenger RNA; they direct the "expression" of the genes.

The chief discoveries of molecular biology took place in a period of only about twenty-five years. Another fifteen years were required before new and more sophisticated technologies, united today under the name of genetic engineering, would permit the isolation and characterization of genes, in particular those of highly complex organisms.

Exploration of the Molecular Dominion

If we evaluate the molecular revolution within the context of biological history, it is easy to note that it is the culmination of a long process which began with the first observations through a microscope. The aim of these early researchers was to understand the functioning of living organisms by describing their organization at the microscopic level. From the end of the 18th century, the characterization of the chemical molecules which make up living beings gained increasingly greater attention, along with the birth of physiological chemistry in the 19th century, developed by the German chemist Justus von Liebig and following the birth of biochemistry at the beginning of the 20th, thanks to another German chemist Eduard Buchner. Between the molecules studied by chemists and the tiny structures visible under the optical microscope, such as the cellular nucleus or the chromosomes, there was an obscure zone, "the world of the ignored dimensions," as it was called by the chemical-physicist Wolfgang Ostwald. This world is populated by colloids, chemical compounds whose structure and properties were not well defined.

The successes of molecular biology derived from the exploration of that unknown world by means of the new technologies developed by chemists and physicists: X-ray diffraction, electron microscopy, ultracentrifugization, and electrophoresis. These studies revealed the structure and function of the macromolecules. A milestone in that process was the work of Dr. Linus Pauling, which for the first time linked the specific genetic mutation in patients with sickle cell disease to a demonstrated change in an individual protein, the hemoglobin in the erythrocytes of heterozygous or homozygous individuals.

Encounter between Biochemistry and Genetics

The development of molecular biology is also the encounter of two disciplines which made considerable progress in the course of the first thirty years of the twentieth century: biochemistry and genetics. The first studies the structure and function of the molecules which make up living things. Between 1900 and 1940, the central processes of metabolism were described: the process of digestion and the absorption of the nutritive elements derived from alimentation, such as the sugars. Every one of these processes is catalyzed by a particular enzyme. Enzymes are proteins, like the antibodies present in blood or the proteins responsible for muscular contraction. As a consequence, the study of proteins, of their structure and synthesis, became one of the principal objectives of biochemists.

The second discipline of biology which developed at the beginning of the 20th century is genetics. After the rediscovery of the laws of Mendel through the studies of Hugo de Vries, Carl Correns and Erich von Tschermak in 1900, this science began to take shape thanks to the adoption by Thomas Hunt Morgan, in 1910, of a model organism for genetic studies, the famous fruit fly (Drosophila melanogaster). Shortly after, Morgan showed that the genes are localized on chromosomes. Following this discovery, he continued working with Drosophila and, along with numerous other research groups, confirmed the importance of the gene in the life and development of organisms. Nevertheless, the chemical nature of genes and their mechanisms of action remained a mystery. Molecular biologists committed themselves to the determination of the structure, and the description of the complex relations between, genes and proteins.

The development of molecular biology was not just the fruit of some sort of intrinsic "necessity" in the history of ideas, but was a characteristically historical phenomenon, with all of its unknowns, imponderables and contingencies: the remarkable developments in physics at the beginning of the 20th century highlighted the relative lateness in development in biology, which became the "new frontier" in the search for knowledge about the empirical world. Moreover, the developments of the theory of information and cybernetics in the 1940s, in response to military exigencies, brought to the new biology a significant number of fertile ideas and, especially, metaphors.

The choice of bacteria and of its virus, the bacteriophage, as models for the study of the fundamental mechanisms of life was almost natural - they are the smallest living organisms known to exist - and at the same time the fruit of individual choices. This model owes its success, above all, to the fame and the sense of organization of Max Delbrück, a German physicist, who was able to create a dynamic research group, based in the United States, whose exclusive scope was the study of the bacteriophage: the *School of the Phage.*

The geographic panorama of the developments of the new biology was conditioned above all by preceding work. The US, where genetics had developed the most rapidly, and the UK, where there was a coexistence of both genetics and biochemical research of highly advanced levels, were in the avant-garde. Germany, the cradle of the revolutions in physics, with the best minds and the most advanced laboratories of genetics in the world, should have had a primary role in the development of molecular biology. But history decided differently: the arrival of the Nazis in 1933 - and, to a less extreme degree, the rigidification of totalitarian measures in fascist Italy - caused the emigration of a large number of Jewish and non-Jewish scientists. The majority of them fled to the US or the UK, providing an extra impulse to the scientific dynamism of those nations. These movements ultimately made molecular biology a truly international science from the very beginnings.

DNA Biochemistry

First Isolation of DNA

Working in the 19th century, biochemists initially isolated DNA and RNA (mixed together) from cell nuclei. They were relatively quick to appreciate the polymeric nature of their "nucleic acid" isolates, but realized only later that nucleotides were of two types—one containing ribose and the other deoxyribose. It was this subsequent discovery that led to the identification and naming of DNA as a substance distinct from RNA.Friedrich Miescher (1844-1895) discovered a substance he called "nuclein" in 1869. Somewhat later, he isolated a pure sample of the material now known as DNA from the sperm of salmon, and in 1889 his pupil, Richard Altmann, named it "nucleic acid". This substance was found to exist only in the chromosomes.

In 1919 Phoebus Levene at the Rockefeller Institute identified the components (the four bases, the sugar and the phosphate chain) and he showed that the components of DNA were linked in the order phosphate-sugar-base. He called each of these units a nucleotide and suggested the DNA molecule consisted of a string of nucleotide units linked together through the phosphate groups, which are the 'backbone' of the molecule. However Levene thought the chain was short and that the bases repeated in the same fixed order. Torbjorn Caspersson and Einar Hammersten showed that DNA was a polymer.

Chromosomes and Inherited Traits

Max Delbrück, Nikolai V. Timofeeff-Ressovsky, and Karl G. Zimmer published results in 1935 suggesting that chromosomes are very large molecules the structure of which can be changed by treatment with X-rays, and that by so changing their structure it was possible to change the heritable characteristics governed by those chromosomes. In 1937 William Astbury produced the first X-ray diffraction patterns from DNA. He was not able to propose the correct structure but the patterns showed that DNA had a regular structure and therefore it might be possible to deduce what this structure was. In 1943, Oswald Theodore Avery and a team of scientists discovered that traits proper to the "smooth" form of the *Pneumococcus* could be transferred to the "rough" form of the same bacteria merely by making the killed "smooth" (S) form available to the live "rough" (R) form. Quite unexpectedly, the living R *Pneumococcus* bacteria were transformed into a new strain of the S form, and the transferred S characteristics turned out to be heritable. Avery called the medium of transfer of traits the transforming principle; he identified DNA as the transforming principle, and not protein as previously thought. He essentially redid Frederick Griffith's experiment. In 1953, Alfred Hershey and Martha Chase did an experiment.

Discovery of the Structure of DNA

In the 1950s, three groups made it their goal to determine the structure of DNA. The first group to start was at King's College London and was led by Maurice Wilkins and was later joined by Rosalind Franklin. Another group consisting of Francis Crick and James D. Watson was at Cambridge. A third group was at Caltech and was led by Linus Pauling. Crick and Watson built physical models using metal rods and balls, in which they incorporated the known chemical structures of the nucleotides, as well as the known position of the linkages joining one nucleotide to the next along the polymer. At King's College Maurice Wilkins and Rosalind Franklin examined X-ray diffraction patterns of DNA fibers. Of the three groups, only the London group was able to produce good quality diffraction patterns and thus produce sufficient quantitative data about the structure.

Helix Structure

In 1948 Pauling discovered that many proteins included helical (see alpha helix) shapes. Pauling had deduced this structure from X-ray patterns and from attempts to physically model the structures. (Pauling was also later to suggest an incorrect three chain helical DNA structure based on Astbury's data.) Even in the initial diffraction data from DNA by Maurice Wilkins, it was evident that the structure involved helices. But this insight was only a beginning. There remained the questions of how many strands came together, whether this number was the same for every helix, whether the bases pointed toward the helical axis or away, and ultimately what were the explicit angles and coordinates of all the bonds and atoms. Such questions motivated the modeling efforts of Watson and Crick.

Complementary Nucleotides

In their modeling, Watson and Crick restricted themselves to what they saw as chemically and biologically reasonable. Still, the breadth of possibilities was very wide. A breakthrough occurred in 1952, when Erwin Chargaff visited Cambridge and inspired Crick with a description of experiments Chargaff had published in 1947. Chargaff had observed that the proportions of the four nucleotides vary between one DNA sample and the next, but that for particular pairs of nucleotides — adenine and thymine, guanine and cytosine — the two nucleotides are always present in equal proportions.

Using X-ray diffraction, as well as other data from Rosalind Franklin and her information that the bases were paired, James D. Watson and Francis Crick arrived at the first accurate model of DNA's molecular structure in 1953, which was accepted through inspection by Rosalind Franklin. The discovery was made on February 28, 1953; the first Watson/Crick paper appeared in Nature on April 25,1953. Sir Lawrence Bragg, the director of the Cavendish Laboratory, where Watson and Crick worked, gave a talk at Guys Hospital Medical School in London on Thursday, May 14, 1953 which resulted in an article by Ritchie Calder in The News Chronicle of London, on Friday, May 15, 1953, entitled "Why You Are You. Nearer Secret of Life." The news reached readers of The New York Times the next day; Victor K. McElheny, in researching his biography, "Watson and DNA: Making a Scientific Revolution", found a clipping of a six-paragraph New York Times article written from London and dated May 16, 1953 with the headline "Form of `Life Unit' in Cell Is Scanned." The article ran in an early edition and was then pulled to make space for news deemed more important.(The New York Times subsequently ran a longer article on June 12, 1953). The Cambridge University undergraduate newspaper also ran its own short article on the discovery on Saturday, May 30th, 1953. Bragg's original announcement at a Solvay conference on proteins in Belgium on 8 April 1953 went unreported by the press! In 1962 Watson, Crick, and Maurice Wilkins jointly received the Nobel Prize for their determination of the structure of DNA.

Central Dogma

Watson and Crick's model attracted great interest immediately upon its presentation.

Arriving at their conclusion on February 21, 1953, Watson and Crick made their first announcement on February 28. In an influential presentation in 1957, Crick laid out the "Central Dogma", which foretold the relationship between DNA, RNA, and proteins, and articulated the "sequence hypothesis." A critical confirmation of the replication mechanism that was implied by the double-helical structure followed in 1958 in the form of the Meselson-Stahl experiment. Work by Crick and coworkers showed that the genetic code was based on non-overlapping triplets of bases, called codons, and Har Gobind Khorana and others deciphered the genetic code not long afterward (1966). These findings represent the birth of molecular biology.

Protein Biochemistry

First Isolation and Classification

Proteins were recognized as a distinct class of biological molecules in the eighteenth century by Antoine Fourcroy and others. Members of this class (called the "albuminoids", *Eiweisskörper*, or *matières albuminoides*) were recognized by their ability to coagulate or flocculate under various treatments such as heat or acid; well-known examples at the start of the nineteenth century included albumen from egg whites, blood serum albumin, fibrin, and wheat gluten. The similarity between the cooking of egg whites and the curdling of milk was recognized even in ancient times; for example, the name *albumen* for the egg-white protein was coined by Pliny the Elder from the Latin *albus ovi* (egg white).

With the advice of Jöns Jakob Berzelius, the Dutch chemist Gerhardus Johannes Mulder carried out elemental analyses of common animal and plant proteins. Mulder published his findings in two papers and hypothesized that there was one basic substance (*Grundstoff*) of proteins, and that it was synthesized by plants and absorbed from them by animals in digestion. Berzelius was an early proponent of this theory and proposed the name "protein" for this substance in a letter dated 10 July 1838

Purifications and measurements of mass

The minimum molecular weight suggested by Mulder's analyses was roughly 9 kDa, hundreds of times larger than other molecules being studied. Hence, the chemical structure of proteins (their primary structure) was an active area of research until 1949, when Fred Sanger sequenced insulin. The (correct) theory that proteins were linear polymers of amino acids linked by peptide bonds was proposed independently and simultaneously by Franz Hofmeister and Emil Fischer at the same conference in 1902. However, some scientists were sceptical that such long macromolecules could be stable in solution. Consequently, numerous alternative theories of the protein primary structure were proposed, e.g., the colloidal hypothesis that proteins were assemblies of small molecules, the cyclol hypothesis of Dorothy Wrinch, the diketopiperazine hypothesis of Emil Abderhalden and the pyrrol/piperidine hypothesis of Troensgard (1942). Most of these theories had difficulties in accounting for the fact that the digestion of proteins yielded peptides and amino acids. Proteins were finally shown to be macromolecules of well-defined composition (and not colloidal mixtures) by Theodor Svedberg using analytical ultracentrifugation. The possibility that some proteins are non-covalent associations of such macromolecules was shown by Gilbert Smithson Adair (by measuring the osmotic pressure of hemoglobin) and, later, by Frederic M. Richards in his studies of ribonuclease S. The mass spectrometry of proteins has long been a useful technique for identifying posttranslational modifications and, more recently, for probing protein structure.

Most proteins are difficult to purify in more than milligram quantities, even using the most modern methods. Hence, early studies focused on proteins that could be purified in large quantities, e.g., those of blood, egg white, various toxins, and digestive/metabolic enzymes obtained from slaughterhouses. Many techniques of protein purification were developed during World War II in a project led by Edwin Joseph Cohn to purify blood proteins to help keep soldiers alive. In the late 1950s, the Armour Hot Dog Co. purified 1 kg (= one million milligrams) of pure bovine pancreatic ribonuclease A and made it freely available to scientists around the world. This generous act made RNase A the main protein for basic research for the next few decades, resulting in several Nobel Prizes.

Protein Folding and First Structural Models

The study of protein folding began in 1910 with a famous paper by Henrietta Chick and C. J. Martin, in which they showed that the flocculation of a protein was composed of two distinct processes: the precipitation of a protein from solution was *preceded* by another process called denaturation, in which the protein became much less soluble, lost its enzymatic activity and became more chemically reactive. In the mid-1920s, Tim Anson and Alfred Mirsky proposed that denaturation was a reversible process, a correct hypothesis that was initially lampooned by some scientists as "unboiling the egg". Anson also suggested that denaturation was a two-state ("all-or-none") process, in which one fundamental molecular transition resulted in the drastic changes in solubility, enzymatic activity and chemical reactivity; he further noted that the free energy changes upon denaturation were much smaller than those typically involved in chemical reactions.

In 1929, Hsien Wu hypothesized that denaturation was protein unfolding, a purely conformational change that resulted in the exposure of amino acid side chains to the solvent. According to this (correct) hypothesis, exposure of aliphatic and reactive side chains to solvent rendered the protein less soluble and more reactive, whereas the loss of a specific conformation caused the loss of enzymatic activity. Although considered plausible, Wu's hypothesis was not immediately accepted, since so little was known of protein structure and enzymology and other factors could account for the changes in solubility, enzymatic activity and chemical reactivity. In the early 1960s, Chris Anfinsen showed that the folding of ribonuclease A was fully reversible with no external cofactors needed, verifying the "thermodynamic hypothesis" of protein folding that the folded state represents the global minimum of free energy for the protein.

The hypothesis of protein folding was followed by research into the physical interactions that stabilize folded protein structures. The crucial role of hydrophobic interactions was hypothesized by Dorothy Wrinch and Irving Langmuir, as a mechanism that might stabilize her cyclol structures. Although supported by J. D. Bernal and others, this (correct) hypothesis was rejected along with the cyclol hypothesis, which was disproven in the 1930s by Linus Pauling (among others). Instead, Pauling championed the idea that protein structure was stabilized mainly by hydrogen bonds, an idea advanced initially by William Astbury.

Remarkably, Pauling's incorrect theory about H-bonds resulted in his *correct* models for the secondary structure elements of proteins, the alpha helix and the beta sheet. The hydrophobic interaction was restored to its correct prominence by a famous article in 1959 by Walter Kauzmann on denaturation, based partly on work by Kaj Linderstrøm-Lang. The ionic nature of proteins was demonstrated by Bjerrum, Weber and Arne Tiselius, but Linderstrom-Lang showed that the charges were generally accessible to solvent and not bound to each other.

The secondary and low-resolution tertiary structure of globular proteins was investigated initially by hydrodynamic methods, such as analytical ultracentrifugation and flow birefringence. Spectroscopic methods to probe protein structure (such as circular dichroism, fluorescence, near-ultraviolet and infrared absorbance) were developed in the 1950s. The first atomic-resolution structures of proteins were solved by X-ray crystallography in the 1960s and by NMR in the 1980s. As of 2006, the Protein Data Bank has nearly 40,000 atomic-resolution structures of proteins. In more recent times, cryo-electron microscopy of large macromolecular assemblies and computational protein structure prediction of small protein domains are two methods approaching atomic resolution.

BASICS OF BIOCHEMISTRY

Biochemistry is the study of the chemical processes in living organisms. It deals with the structure and function of cellular components such as proteins, carbohydrates, lipids, nucleic acids and other biomolecules. Although there are a vast number of different biomolecules many are complex and large molecules (called *polymers*) that are composed of similar repeating subunits (called *monomers*). Each class of polymeric biomolecule has a different set of subunit types. For example, a protein is a polymer whose subunits are selected from a set of 20 or more amino acids. Biochemistry studies the chemical properties of important biological molecules, like proteins, in particular the chemistry of enzyme-catalyzed reactions.

The biochemistry of cell metabolism and the endocrine system has been extensively described. Other areas of biochemistry include the genetic code (DNA, RNA), protein synthesis, cell membrane transport, and signal transduction. Since all known life forms that are still alive today are descended from one of two common ancestor, they have generally similar biochemistries. It is remaining unknown whether alternative types of biochemistries are possible, practical, or not; with the elements of the matter which our Universe is made by. An emerging thesis considers that only carbon-based compounds are available to be part of a real biochemistry; this thesis is called "carbon chauvinism".Originally, it was generally believed that life was not subject to the laws of science the way non-life was. It was thought that only living beings could produce the molecules of life (from other, previously existing biomolecules). Then, in 1828, Friedrich Wöhler published a paper on the synthesis of urea, proving that organic compounds can be created artificially.

The dawn of biochemistry may have been the discovery of the first enzyme, diastase (today called amylase), in 1833 by Anselme Payen. Eduard Buchner contributed the first demonstration of a complex biochemical process outside of a cell in 1896: alcoholic fermentation in cell extracts of yeast. Although the term "biochemistry" seems to have been first used in 1882, it is generally accepted that the formal coinage of biochemistry occurred in 1903 by Carl Neuberg, a German chemist.

Previously, this area would have been referred to as physiological chemistry. Since then, biochemistry has advanced, especially since the mid-20th century, with the development of new techniques such as chromatography, X-ray diffraction, dual polarisation interferometry, NMR spectroscopy, radioisotopic labeling, electron microscopy and molecular dynamics simulations. These techniques allowed for the discovery and detailed analysis of many molecules and metabolic pathways of the cell, such as glycolysis and the Krebs cycle (citric acid cycle).Another significant historic event in biochemistry is the discovery of the gene and its role in the transfer of information in the cell. This part of biochemistry is often called molecular biology. In the 1950s, James D. Watson, Francis Crick, Rosalind Franklin, and Maurice Wilkins were instrumental in solving DNA structure and suggesting its relationship with genetic transfer of information. In 1958, George Beadle and Edward Tatum received the Nobel Prize for work in fungi showing that one gene produces one enzyme. In 1988, Colin Pitchfork was the first person convicted of murder with DNA evidence, which led to growth of forensic science. More recently, Andrew Z. Fire and Craig C. Mello received the 2006 Nobel Prize for discovering the role of RNA interference (RNAi), in the silencing of gene expression.

Today, there are three main types of biochemistry. Plant biochemistry involves the study of the biochemistry of autotrophic organisms such as photosynthesis and other plant specific biochemical processes. General biochemistry encompasses both plant and animal biochemistry. Human/medical/medicinal biochemistry focuses on the biochemistry of humans and medical illnesses.

Biochemistry is a hybrid science: Biology is the science of living organisms and chemistry is the science of atoms and molecules, so biochemistry is the science of the atoms and molecules in living organisms. Its domain encompasses all the living world with the unifying interest in the chemical structures and reactions that occur in living systems.

Scientists use biochemistry today to explain how our body works. We try to make medicines that affect the body a certain way, try to crack the DNA code, try to eat the right foods, explain diseases, observe chemical reactions, and many more using biochemistry.

It can be divided into three principal areas:

1. The structural chemistry of the components of living matter and the relationship of biological function to chemical structure.
2. Metabolism - the totality of chemical reactions that occur in living matter.
3. The chemistry of processes and substances that store and transmit biological information.

Biochemistry's roots as a distinct field of study date to the early 19th century, with the pioneering work of Friedrich Wöhler. Prior to that time, it was believed that the substances in living matter were somehow qualitatively different from those in nonliving matter and did not behave according to the known laws of physics and chemistry. In 1828 Wöhler showed that urea, a substance of biological origin, could be synthesized in the laboratory from the inorganic compound ammonium cyanate.

Chromosomes were discovered in 1875 by Walter Flemming and identified as genetic elements by 1902. The development of the electron microscope, between about 1930 and 1950, provided a whole new level of insight into cellular structure. With it, subcellular organelles like mitochondria and chloroplasts could be studied, and it was realised that specific biochemical processes were localised in these subcellular particles.

Nucleic acids had been isolated in 1869 by Friedrich Miescher, but their chemical structures were poorly understood, and in the early 1900s they were thought to be simple substances, fit only for structural roles in the cell. The idea of the gene, a unit of hereditary information, was first proposed in the mid-nineteenth century by Gregor Mendel. By about 1900, cell biologists realised that genes must be found in chromosomes, which are composed of proteins and nucleic acids. Most biochemists believed that only the proteins were structurally complex enough to carry genetic information.

That belief was dead wrong. Experiments in the 1940s and early 1950s proved conclusively that deoxyribonucleic acid (DNA) is the bearer of genetic information. One of the most important advances in the history of science occurred in 1953, when James Watson and Francis Crick described the double-helical structure of DNA. This concept immediately suggested ways in which information could be encoded in the structure of molecules and transmitted intact from one generation to the next.

In order to understand the impact of biochemistry on biology, one must understand the chemical elements of living matter and the complete structures of many biological compounds - amino acids, sugars, lipids, nucleotides, vitamins, and hormones - and their behaviour during metabolic reactions. Essential understanding of biochemistry requires knowledge of the stoichiometry and mechanisms of a large number of reactions. In addition, an understanding of the basic thermodynamic principles is essential for learning how plants derive energy from sunlight (photosynthesis) and how animals derive energy from food (catabolism).

Living creatures on the earth are composed mainly of a very few elements, principally carbon, hydrogen, oxygen, and nitrogen (C, H, O, N). Life is not built on these four elements alone. Many other elements are necessary for terrestrial organisms . A "second tier" of essential elements includes sulfur and phosphorus, which form covalent bonds, and the ions Na^+, K^+, Mg^{2+}, Ca^{2+}, and Cl^-.

Sulfur is an important constituent of proteins, and phosphorus plays essential roles in energy metabolism and the structure of nucleic acids. Beyond the first two tiers of elements, we come to those that play quantitatively minor, but often indispensable, roles. Most of these third- and fourth-tier elements are metals, and some serve as aids to catalysis of biochemical reactions.

Biological Molecules

The complexity of life processes requires that many of the molecules that participate in these processes be enormous. The most extreme example is DNA. Consider, for instance, the DNA molecules released from one human chromosome. The long, looped thread corresponds to just two enormous molecules, each with a molecular weight of about 20 billion. Protein molecules are generally much smaller, but they are still large, with a typical protein having a mass of 50,000 Daltons.

All large macromolecules in the cell are polymers, made by joining prefabricated units, or monomers. The monomers of a given type of macromolecule are of limited diversity and are linked together, or polymerised, by identical mechanisms. A simple example is the carbohydrate cellulose, a major constituent of the cell walls of plants. Cellulose is a polymer made by joining thousands of molecules of glucose, a simple sugar. In this polymer all of the chemical linkages between the monomers are identical. Covalent links between glucose units are formed by removing a water molecule between two adjoining molecules; the portion of a glucose molecule remaining in the chain is called a glucose residue. Because cellulose is a polymer of a simple sugar, or saccharide, it is called a polysaccharide. This particular polymer is constructed from identical monomeric units, so it is called a homopolymer. In contrast, many polysaccharides and all nucleic acids and proteins are heteropolymers, polymers constructed from a number of different kinds of monomer units. Nucleic acids are polymers of four nucleotides, so nucleic acids are also called polynucleotides. Similarly, proteins are assembled from combinations of 20 different amino acids. Protein chains are called polypeptides, a term derived from the peptide bond that joins two amino acids together.

Amino Acids

Amino acids are organic acids containing an amine group. The most common amino acids are -amino acids and the most common α–amino acids are the L—amino acids.It is important to understand the following about amino acid structure:

1. Only 20 L—amino acids are used to make proteins. Rare exceptions are bacterial membrane proteins, which contain a few D-Amino Acids, modified amino acids - primarily lysine and proline, and occasional incorporation of the rare amino acid, selenocysteine).
2. Side groups are what distinguish the α-amino acids from each other.
3. Amino acids can exist as zwitterions - substances containing equal numbers of positive and negative charge - due to their carboxyl and amine groups, which can be negatively and positively charged, respectively.

Several common amino acids found in cells, such as ornithine and citrulline, are not used to make proteins. Amino acids are the monomeric units or "building blocks" of proteins that are joined together covalently in peptide bonds. Amino acids are built into proteins by the process of translation using the genetic code.

Sugars

Sugars and polymers of them are known as carbohydrates because their general formulas can all be written in a simple form as $(CH_2O)x$, as if they were a hydrated form of carbon. Another term used to describe sugar-based molecules is rooted in the word "saccharide" (from the Latin, saccharum, meaning sugar). Single units of sugar, such as glucose, are called monosaccharides. Sugars can be readily linked together to form disaccharides (contains two sugars), oligosaccharides and polysaccharides

Lipids

Lipids are a class of nonpolar molecules that include the fats, cholesterol, fatty acids, lipid-soluble vitamins, waxes, soaps, glycerophospholipids, sphingolipids, and others. Lipids are found in the membranes of cells, the endoplasmic reticulum, and in specialised fat storage cells called adipocytes.

Nucleotides

Ribonucleotides and deoxyribonucleotides (collectively called nucleotides) are the building blocks of the nucleic acids, RNA and DNA, respectively. Nucleotides are composed of a sugar, a phosphate, and a base. The composition of DNA differs from that of RNA in only two respects. First, DNA uses nucleotides containing the sugar deoxyribose. RNA's nucleotides contain the sugar ribose. Second, DNA uses the bases adenine, guanine, cytosine, and thymine. RNA uses the bases adenine, guanine, cytosine, and uracil Nucleotides differ from nucleosides in that the latter do not contain phosphate, so we sometimes refer to nucleotides as nucleoside (mono,di, or tri)-phosphates. For example, adenosine diphosphate is a nucleotide (also called ADP) Deoxyribonucleotides (written with a 'd') differ from ribonucleotides in containing deoxyribose as the sugar moiety instead of ribose.

Vitamins

Vitamins are organic molecules that are essential to the biological processes of higher organisms, but cannot be synthesized by these organisms. Examples include the fat-soluble vitamins A, D, E, and K and the water-soluble B complex vitamins and ascorbic acid, also known as vitamin C.

Hormones

A hormone is a substance that is synthesized and secreted by specialised cells and carried via the circulation to target cells, where it elicits specific changes in the metabolic behavior of the cell by interacting with a hormone-specific receptor.Chemically, the hormones in vertebrate metabolism include:

(1) peptides or polypeptides, such as insulin or glucagon

(2) steroid, including glucocorticoids and the sex hormones

(3) amino acid derivatives, including the catecholamines and thyroxine.

Uses of Biochemistry

The results of biochemical research are used extensively in the world outside the laboratory - in agriculture, medical sciences, nutrition, and many other fields. In clinical chemistry, biochemical measurements on people help diagnose illnesses and monitor responses to treatment. Liver disease is now routinely diagnosed and monitored by measurements of blood levels of enzymes called transaminases and of a hemoglobin breakdown product called bilirubin.Pharmacology and toxicology are concerned with the effects of external chemical substances on metabolism. Drugs and poisons usually act by interfering with specific metabolic pathways. A good example is the antibiotic penicillin, which kills bacteria by inhibiting an enzyme that synthesizes an essential polysaccharide of the bacterial cell wall. Because animal cells do not synthesize these polysaccharides, they are not harmed by this inhibitor, which can therefore be used therapeutically.

A particularly exciting prospect in contemporary biochemistry is that of creating so-called designer drugs. If the target site for action of a drug is a protein enzyme or receptor, determining the detailed molecular structure of that target allows us to design inhibitors that bind to it with great selectivity. Early products of this drug architecture are now being tested. In the mid-1980s the detailed molecular structures of viruses began to be determined, opening the possibility of rational design of antiviral agents as well.

Herbicides and pesticides, in many instances, act in similar ways - by blocking enzymes or receptors in the target organism. The first generations of these toxic agents (such as DDT) were so nonspecific in their effects that organisms other than the target populations were often affected, resulting in unforeseen and often severe damage to the environment. Furthermore, the indiscriminate use of these agents gave rise to resistant target populations, so that an ever-increasing number of toxins needed to be used. Biochemistry is involved in understanding the actions of herbicides and pesticides, in increasing their selectivity, and in understanding and dealing with mechanisms by which target organisms become resistant to them. Thus, biochemistry has become an important component of environmental science.

References

Boyer, Rodney, *Concepts in Biochemistry,* Pacific Grove, CA: Brooks/Cole Publishing Company, 1999.

Jorde, L.B., J.C. Carey, M.J. Bamshad, and R.L. White, *Medical Genetics,* 2nd ed. St. Louis: Mosby-Year Book, Inc., 2000.

Lodish, H., et. al., *Molecular Cell Biology,* 4th ed. New York: W. H. Freeman & Co., 2000.

Nelson, David L., David L. Nelson, and Michael M. Cox, *Lehninger Principles of Biochemistry,* 3rd edition. Worth Publishing, 2000.

Fruton, Joseph. *Proteins, Genes, Enzymes: The Interplay of Chemistry and Biology*. New Haven: Yale University Press. 1999. ISBN 0-300-07608-8

Lily E. Kay, *The Molecular Vision of Life: Caltech, the Rockefeller Foundation, and the Rise of the New Biology,* Oxford University Press, Reprint 1996

Morange, Michel. *A History of Molecular Biology*. Cambridge, MA: Harvard University Press. 1998.

2

Chemistry of Biomolecules

Biochemistry is the study of the chemistry related to biological organisms. It forms a bridge between biology and chemistry by studying how complex chemical reactions and chemical structures give rise to life and life's processes. Biochemistry is sometimes viewed as a hybrid branch of organic chemistry which specializes in the chemical processes and chemical transformations that take place inside of living organisms, but the truth is that the study of biochemistry should generally be considered neither fully "biology" nor fully "chemistry" in nature. Biochemistry incorporates everything in size between a molecule and a cell and all the interactions between them. Biochemistry essentially remains the study of the structure and function of cellular components and the processes carried out both on and by organic macromolecules(as proteins, carbohydrates, lipids, nucleic acids, and other biomolecules). All life forms alive today are believed to have descended from a single ancestor, because all known living beings have similar biochemistries.

In order to understand the impact of biochemistry on biology, one must understand the chemical elements of living matter and the complete structures of many biological compounds - amino acids, sugars, lipids, nucleotides, vitamins, and hormones - and their behaviour during metabolic reactions. Essential understanding of biochemistry requires knowledge of the stoichiometry and mechanisms of a large number of reactions. In addition, an understanding of the basic thermodynamic principles is essential for learning how plants derive energy from sunlight (photosynthesis) and how animals derive energy from food (catabolism).Living creatures on the earth are composed mainly of a very few elements, principally carbon, hydrogen, oxygen, and nitrogen (C, H, O, N). Life is not built on these four elements alone. Many other elements are necessary for terrestrial organisms . A "second tier" of essential elements includes sulfur and phosphorus, which form covalent bonds, and the ions Na^+, K^+, Mg^{2+}, Ca^{2+}, and Cl^-. Sulfur is an important constituent of proteins, and phosphorus plays essential roles in energy metabolism and the structure of nucleic acids. Beyond the first two tiers of elements, we come to those that play quantitatively minor, but often indispensable, roles. Most of these third- and fourth-tier elements are metals, and some serve as aids to catalysis of biochemical reactions.

NATURE AND TYPES OF BIOMOLECULES

A biomolecule is any organic molecule that is produced by a living organism, including large polymeric molecules such as proteins, polysaccharides, and nucleic acids as well as small molecules such as primary metabolites, secondary metabolites, and natural products. As organic molecules, biomolecules consist primarily of carbon and hydrogen, nitrogen, and oxygen, and, to a smaller extent, phosphorus and sulfur. Other elements sometimes are incorporated but are much less common.The complexity of life processes requires that many of the molecules that participate in these processes be enormous. The most extreme example is DNA. Consider, for instance, the DNA molecules released from one human chromosome. The long, looped thread corresponds to just two enormous molecules, each with a molecular weight of about 20 billion. Protein molecules are generally much smaller, but they are still large, with a typical protein having a mass of 50,000 Daltons.

All large macromolecules in the cell are polymers, made by joining prefabricated units, or monomers. The monomers of a given type of macromolecule are of limited diversity and are linked together, or polymerised, by identical mechanisms. A simple example is the carbohydrate cellulose, a major constituent of the cell walls of plants. Cellulose is a polymer made by joining thousands of molecules of glucose, a simple sugar. In this polymer all of the chemical linkages between the monomers are identical. Covalent links between glucose units are formed by removing a water molecule between two adjoining molecules; the portion of a glucose molecule remaining in the chain is called a glucose residue.Because cellulose is a polymer of a simple sugar, or saccharide, it is called a polysaccharide. This particular polymer is constructed from identical monomeric units, so it is called a homopolymer. In contrast, many polysaccharides and all nucleic acids and proteins are heteropolymers, polymers constructed from a number of different kinds of monomer units. Nucleic acids are polymers of four nucleotides, so nucleic acids are also called polynucleotides. Similarly, proteins are assembled from combinations of 20 different amino acids. Protein chains are called polypeptides, a term derived from the peptide bond that joins two amino acids together.

A diverse range of biomolecules exist, including:

1. Small molecules:
   - Lipid, phospholipid, glycolipid, sterol, glycerolipid
   - Vitamin
   - Hormone, neurotransmitter
   - Carbohydrate, sugar
2. Monomers:
   - Amino acids

- Nucleotides
- Monosaccharides

3. Polymers:
   - Peptides, oligopeptides, polypeptides, proteins
   - Nucleic acids, i.e. DNA, RNA
   - Oligosaccharides, polysaccharides (including cellulose)
   - Lignin

AMINO ACIDS

Amino acids contain both amino and carboxylic acid functional groups. (In biochemistry, the term amino acid is used when referring to those amino acids in which the amino and carboxylate functionalities are attached to the same carbon, plus proline which is not actually an amino acid). Amino acids are the building blocks of long polymer chains. With 2-10 amino acids such chains are called peptides, with 10-100 they are often called polypeptides, and longer chains are known as proteins. These protein structures have many structural and enzymatic roles in organisms. There are twenty amino acids that are encoded by the standard genetic code, but there are more than 500 natural amino acids. When amino acids other than the set of twenty are observed in proteins, this is usually the result of modification after translation (protein synthesis).

Amino acids are used for a variety of purposes. The food industry requires Lglutamate as a flavour enhancer, or glycine as sweetener in juices, for instance. The pharmaceutical industry requires the amino acids themselves in infusions—in particular the essential amino acids—or in special dietary foods. And last, but not least, a large market for amino acids is their use as feed additives. The reason is that typical animal feed, like soybean meal for pigs, is poor in some essential amino acids, like methionine and lysine.

Amino acids are molecules containing both amine and carboxyl functional groups. These molecules are particularly important in biochemistry, where this term refers to alpha-amino acids with the general formula $H_2NCHRCOOH$, where R is an organic substituent.In the alpha amino acids, the amino and carboxylate groups are attached to the same carbon atom, which is called the a–carbon. The various alpha amino acids differ in which side chain (R group) is attached to their alpha carbon. They can vary in size from just a hydrogen atom in glycine through a methyl group in alanine to a large heterocyclic group in tryptophan.

Amino acids are critical to life, and have a variety of roles in metabolism. One particularly important function is as the building blocks of proteins, which are linear chains of amino acids. Amino acids are also important in many other biological molecules, such as forming parts of coenzymes, as in S-adenosylmethionine, or as precursors for the biosynthesis of molecules such as heme. Due to this central role in biochemistry, amino acids are very important in nutrition.Amino acids are commonly used in food technology and industry. For example, monosodium glutamate is a common flavor enhancer that gives foods the taste called umami. Beyond the amino acids that are found in all forms of life, amino acids are also used in industry. Applications include the production of biodegradable plastics, drugs and chiral catalysts.

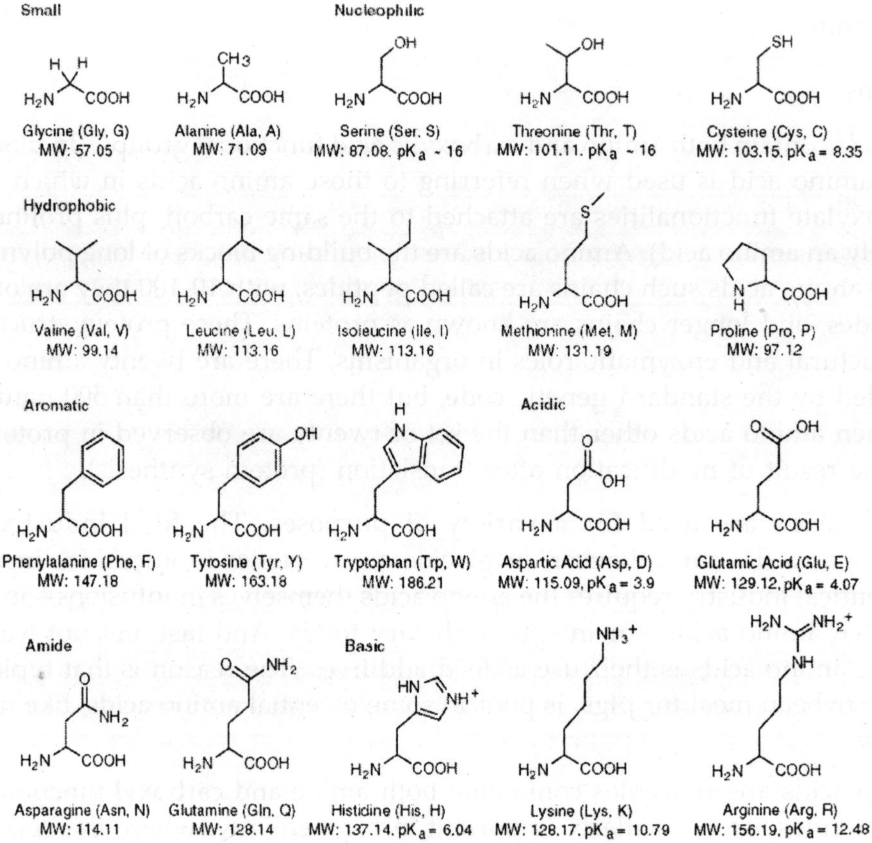

Figure 1. Amino acids

Alpha-amino acids are the building blocks of proteins. Amino acids combine in a condensation reaction, that is, through dehydration synthesis, that releases water and the new "amino acid residue" that is held together by a peptide bond. Proteins are defined by their unique sequence of amino acid residues; this sequence is the primary structure of the protein. Just as the letters of the alphabet can be combined to form an almost endless variety of words, amino acids can be linked in varying sequences to form a vast variety of proteins.Twenty-two standard amino acids are used by cells in protein biosynthesis, and these are specified by the general genetic code.These twenty-two amino acids are biosynthesized from other molecules, but organisms differ in which ones they can synthesize and which ones must be provided in their diet. The ones that cannot be synthesized by an organism are called essential amino acids.

General Structure

In the structure shown at the top of the page, **R** represents a side chain specific to each amino acid. The carbon atom next to the carbonyl group is called the α–carbon and amino acids with a side chain bonded to this carbon are referred to as *alpha amino acids*. These are the most common form found in nature. In the alpha amino acids, the α–carbon is a chiral carbon atom, with the exception of glycine. In amino acids that have a carbon chain attached to the α–carbon (such as lysine, shown to the right) the carbons are labeled in order as α, β, γ, δ, and so on. In some amino acids, the amine group is attached to the β or γ-carbon, and these are therefore referred to as *beta* or *gamma amino acids*.

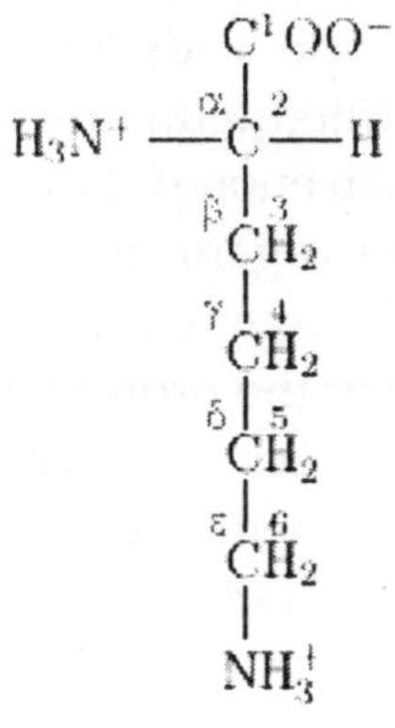

Figure 2. General Structure of Amino acids

Amino acids are usually classified by the properties of their side chain into four groups. The side chain can make an amino acid a weak acid or a weak base, and a hydrophile if the side chain is polar or a hydrophobe if it is nonpolar. The chemical structures of the twenty-two standard amino acids, along with their chemical properties, are described more fully in the article on these proteinogenic amino acids.

The phrase "branched-chain amino acids" or BCAA is sometimes used to refer to the amino acids having aliphatic side chains that are non-linear; these are leucine, isoleucine, and valine. Proline is the only proteinogenic amino acid whose side group links to the a-amino group and, thus, is also the only proteinogenic amino acid containing a secondary amine at this position.Chemically, proline is an imino acid since it lacks a primary amino group, but it is classed as an amino acid in the current biochemical nomenclature, and may also be called an "N-alkylated alpha-amino acid".

Proteins

Proteins are the major components of living organisms and perform a wide range of essential functions in cells. While DNA is the information molecule, it is proteins that do the work of all cells - microbial, plant, animal. Proteins regulate metabolic activity, catalyse biochemical reactions and maintain structural integrity of cells and organisms. Proteins can be classified in a variety of ways, including their biological function.Proteins can be formed using 20 different building blocks called amino acids. Each of these amino acid building blocks has a different chemical structure and different properties. Each protein has a unique amino acid sequence that is genetically determined by the order of nucleotide bases in the DNA, the genetic code. Since each protein has different numbers and kinds of the twenty available amino acids, each protein has a unique chemical composition and structure. For example, two proteins may each have 37 amino acids but if the sequence of the amino acids is different, then the protein will be different. How many different proteins can be formed from the twenty different amino acids? Consider a protein containing 100 different amino acids linked into one chain. Since each of the 100 positions of this chain could be filled with any one of the 20 amino acids, there are 20100 possible combinations, more than enough to account for the 90-100 million different proteins that may be found in higher organisms. A change in just one amino acid can change the structure and function of a protein. For example, sickle cell anemia is a disease that results from an altered structure of the protein hemoglobin, resulting from a change of the sixth amino acid from glutamic acid to valine. This single amino acid change is enough to change the conformation of hemoglobin so that this protein clumps at lower oxygen concentrations and causes the characteristic sickle shaped red blood cells of the disease. The unique structure and chemical composition of each protein is important for its function; it is also important for separating proteins in a protein purification strategy. Each of these differences in properties can be used as a basis for the separation methods that are used to purify proteins. Because these differences in protein properties originate from differences in the chemical structure of the amino acids that make up the protein.

Structure of Proteins

Amino acids are composed of carbon, hydrogen, oxygen, and nitrogen. Two amino acids, cysteine and methionine, also contain sulfur. The generic form of an amino acid is shown in Figure 3. Atoms of these elements are arranged into 20 kinds of amino acids that are commonly found in proteins. All proteins in all species, from bacteria to humans, are constructed from the same set of twenty amino acids. All amino acids have an amino group (NH_2) and a carboxyl group (COOH) bonded to the same carbon atom, known as the alpha carbon. Amino acids differ in the side chain or R group that is bonded to the alpha carbon. Glycine, the simplest amino acid has a single hydrogen atom as its R group - Alanine has a methyl ($-CH_3$) group.

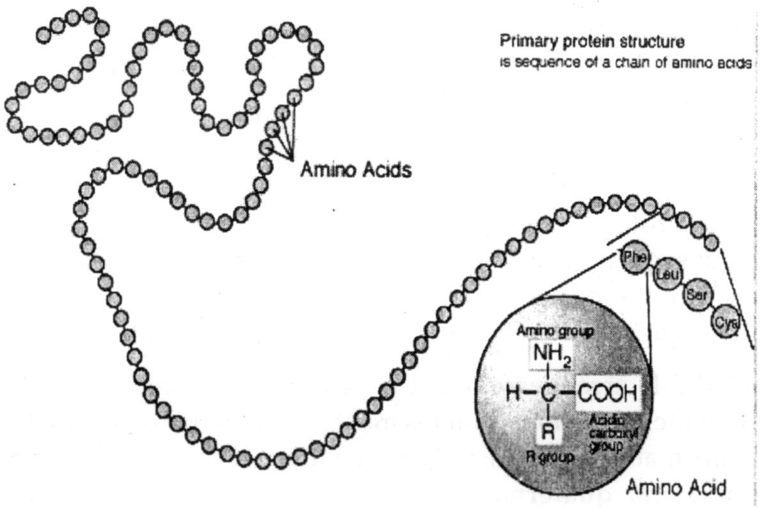

Figure 3. A diagram showing a generic amino acid structure

The chemical composition of the unique R groups is responsible for the important characteristics of amino acids such as chemical reactivity, ionic charge and relative hydrophobicity. The amino acids are grouped according to their polarity and charge. They are divided into four categories, those with polar uncharged R groups, those with apolar (nonpolar) R groups, acidic (charged) and basic (charged) groups. The twenty amino acids, grouped according to the character of their side chain or R group. The polar amino acids are soluble in water because their R groups can form hydrogen bonds with water. For example, serine, threonine and tyrosine all have hydroxyl groups (OH). Amino acids that carry a net negative charge at neutral pH contain a second carboxyl group. These are the acidic amino acids, aspartic acid and glutamic acid, also called aspartate and glutamate, respectively. The basic amino acids have R groups with a net positive charge at pH 7.0. These include lysine, arginine and histidine. There are eight amino acids with nonpolar R groups. As a group, these amino acids are less soluble in water than the polar amino acids. If a protein has a greater percentage of nonpolar R groups, the protein will be more hydrophobic (water hating) in character.

A protein is formed by amino acid subunits linked together in a chain. The bond between two amino acids is formed by the removal of a H_2O molecule from two different amino acids, forming a dipeptide. The bond between two amino acids is called a peptide bond and the chain of amino acids is called a peptide (20 amino acids or smaller) or a polypeptide.

Each protein consists of one or more unique polypeptide chains. Most proteins do not remain as linear sequences of amino acids; rather, the polypeptide chain undergoes a folding process. The process of protein folding is driven by thermodynamic considerations. This means that each protein folds into a configuration that is the most stable for its particular chemical structure and its particular environment. The final shape will vary but the majority of proteins assume a globular configuration.

Many proteins such as myoglobin consist of a single polypeptide chain; others contain two or more chains. For example, hemoglobin is made up of two chains of one type (amino acid sequence) and two of another type.Although the primary amino acid sequence determines how the protein folds, this process is not completely understood. Although certain amino acid sequences can be identified as more likely to form a particular conformation, it is still not possible to completely predict how a protein will fold based on its amino acid sequence alone, and this is an active area of biochemical research.

The final folded 3-D arrangement of the protein is referred to as its conformation. In order to maintain their function, proteins must maintain this conformation. To describe this complex conformation, scientists describe four levels of organisation: primary, secondary, tertiary, and quaternary. The overall conformation of a protein is the combination of its primary, secondary, tertiary and quaternary elements.

Primary Structure refers to the linear sequence of amino acids that make up the polypeptide chain. This sequence is determined by the genetic code, the sequence of nucleotide bases in the DNA. The bond between two amino acids is a peptide bond. This bond is formed by the removal of a H20 molecule from two different amino acids, forming a dipeptide. The sequence of amino acids determines the positioning of the different R groups relative to each other. This positioning therefore determines the way that the protein folds and the final structure of the molecule.

The secondary structure of protein molecules refers to the formation of a regular pattern of twists or kinks of the polypeptide chain. The regularity is due to hydrogen bonds forming between the atoms of the amino acid backbone of the polypeptide chain. The two most common types of secondary structure are called the alpha helix and β pleated sheet. (Figure 4)

Tertiary structure refers to the three dimensional globular structure formed by bending and twisting of the polypeptide chain. This process often means that the linear sequence of amino acids is folded into a compact globular structure. The folding of the polypeptide chain is stabilised by multiple weak, noncovalent interactions. These interactions include:

- Hydrogen bonds that form when a Hydrogen atom is shared by two other atoms.
- Electrostatic interactions that occur between charged amino acid side chains. Electrostatic interactions are attractions between positive and negative sites on macromolecules.

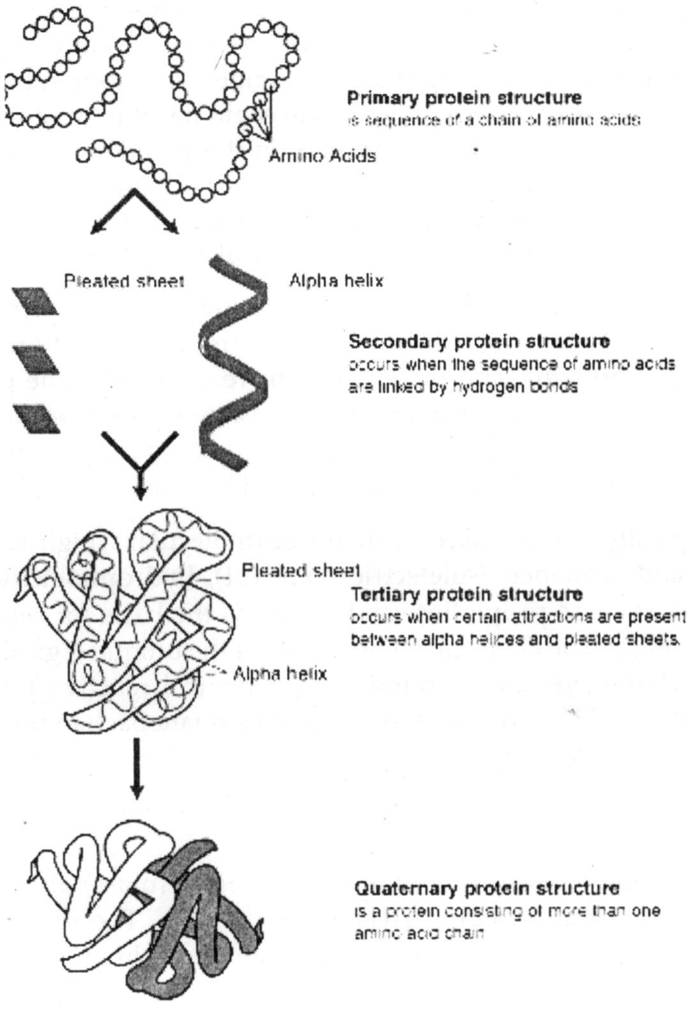

Figure 4. Four levels of protein structure

During folding of the polypeptide chain, amino acids with a polar (water soluble) side chain are often found on the surface of the molecule while amino acids with non polar (water insoluble) side chain are buried in the interior. This means that the folded protein is soluble in water or aqueous solutions. Covalent bonds may also contribute to tertiary structure. The amino acid, cysteine, has an SH group as part of its R group and therefore, the disulfide bond (S-S) can form with an adjacent cysteine. For example, insulin has two polypeptide chains that are joined by two disulfide bonds.

Quaternary structure refers to the fact that some proteins contain more than one polypeptide chain, adding an additional level of structural organisation: the association of the polypeptide chains. Each polypeptide chain in the protein is called a subunit. The subunits can be the same polypeptide chain or different ones. For example, the enzyme b-galactosidase is a tetramer, meaning that it is composed of four subunits, and, in this case, the subunits are identical - each polypeptide chain has the same sequence of amino acids. Hemoglobin, the oxygen carrying protein in the blood, is also a tetramer but it is composed of two polypeptide chains of one type (141 amino acids) and two of a different type (146 amino acids). In chemical shorthand, this is referred to as a2b2 . For some proteins, quaternary structure is required for full activity (function) of the protein.

Some proteins combine with other kinds of molecules such as carbohydrates, lipids, iron and other metals, or nucleic acids, to form glycoproteins, lipoproteins, hemoproteins, metalloproteins, and nucleoproteins respectively. The presence of these other biomolecules affects the protein properties. For example, a protein that is conjugated to carbohydrate, called a glycoprotein, would be more hydrophilic in character while a protein conjugated to a lipid would be more hydrophobic in character.

Proteins are typically characterised by their size (molecular weight) and shape, amino acid composition and sequence, isolelectric point (pI), hydrophobicity, and biological affinity. Differences in these properties can be used as the basis for separation methods in a purification strategy. The chemical composition of the unique R groups is responsible for the important characteristics of amino acids, chemical reactivity, ionic charge and relative hydrophobicity. Therefore protein properties relate back to number and type of amino acids that make up the protein.

Size of proteins

Size of proteins is usually measured in molecular weight (mass) although occasionally the length or diameter of a protein is given in Angstroms. The molecular weight of a protein is the mass of one mole of protein, usually measured in units called daltons. One dalton is the atomic mass of one proton or neutron. The molecular weight can be estimated by a number of different methods including electrophoresis, gel filtration, and more recently by mass spectrometry. The molecular weight of proteins varies over a wide range. For example, insulin is 5,700 daltons while snail hemocyanin is 6,700,000 daltons. The average molecular weight of a protein is between 40,000 to 50,000 daltons. Molecular weights are commonly reported in kilodaltons or (kD), a unit of mass equal to 1000 daltons. Most proteins have a mass between 10 and 100 kD. A small protein consists of about 50 amino acids while larger proteins may contain 3,000 amino acids or more. One of the larger amino acid chains is myosin, found in muscles, which has 1,750 amino acids.

Separation methods that are based on size and shape include gel filtration chromatography (size exclusion chromatography) and polyacrylamide gel electrophoresis.

Amino Acid Composition and Sequence

The amino acid composition is the percentage of the constituent amino acids in a particular protein while the sequence is the order in which the amino acids are arranged.

Each protein has an amino group at one end and a carboxyl group at the other end as well as numerous amino acid side chains, some of which are charged. Therefore each protein carries a net charge. The net protein charge is strongly influenced by the pH of the solution. To explain this phenomenon, consider the hypothetical protein in Figure 5. At pH 6.8, this protein has an equal number of positive and negative charges and so there is no net charge on the protein.

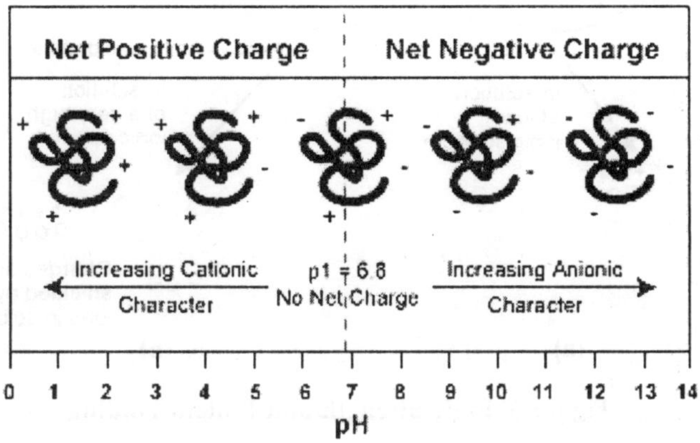

Figure 5. The pI is the pH at which there is no net charge on the protein.

As the pH drops, more H^+ ions are available in the solution. These hydrogen ions bind to negative sites on the amino acids. Therefore, as the pH drops, the protein as a whole becomes positively charged. Conversely, at a basic pH, the protein becomes negatively charged. pH 6.8 is called the pI, or isoelectric point, for this protein; that is, the pH at which there are an equal number of positive and negative charges. Different proteins have different numbers of each of the amino acid side chains and therefore have different isoelectric points. So, in a buffer solution at a particular pH, some proteins will be positively charged, some proteins will be negatively charged and some will have no charge. Separation techniques that are based on charge include ion exchange chromatography, isoelectric focusing and chromato-focusing.

Literally, hydrophobic means fear of water. In aqueous solutions, proteins tend to fold so that areas of the protein with hydrophobic regions are located in internal surfaces next to each other and away from the polar water molecules of the solution. Polar groups on the amino acid are called hydrophilic (water loving) because they will form hydrogen bonds with water molecules. The number, type and distribution of nonpolar amino acid residues within the protein determines its hydrophobic character. A separation method that is based on the hydrophobic character of proteins is hydrophobic interaction chromatography.

Cytoplasmic proteins have mostly hydrophilic (polar) amino acids on their surface and are therefore water soluble, with more hydrophobic groups located on the interior of the protein, sheltered from the aqueous environment. In contrast, proteins that reside in the lipid environment of the cell membrane have mostly hydrophobic amino acids (non polar) on their exterior surface and are not readily soluble in aqueous solutions.

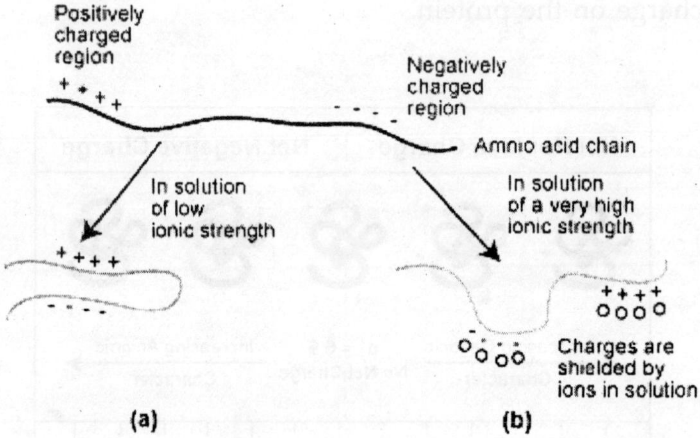

Figure 6. Ionic Strength and Protein Folding

Each protein has a distinct and characteristic solubility in a defined environment and any changes to those conditions (buffer or solvent type, pH, ionic strength, temperature, etc.) can cause proteins to lose the property of solubility and precipitate out of solution. The environment can be manipulated to bring about a separation of proteins- for example, the ionic strength of the solution can be increased or decreased, which will change the solubility of some proteins.

Biological Functions of Proteins

Proteins often interact with other molecules in vivo in a specific way- in other words, they have a biological affinity for that molecule. These molecular counterparts, termed ligands, can be used as "bait" to "fish" out the target protein that you want to purify. For example, one such molecular pair is insulin and the insulin receptor. If you want to purify (or catch) the insulin receptor, you could couple many insulin molecules to a solid support and then run an extract (containing the receptor) over that column. The receptor would be "caught" by the insulin bait. These specific interactions are often exploited in protein purification procedures. Affinity chromatography is a very common method for purifying recombinant proteins (proteins produced by genetic engineering).

Several histidine residues can be engineered at the end of a polypeptide chain. Since repeated histidines have an affinity for metals, a column of the metal can be used as bait to "catch" the recombinant protein.

NUCLEOTIDES

Nucleotides are ubiquitous molecules with considerable structural diversity. There are eight common varieties of nucleotides, each composed of a nitrogenous base linked to a sugar to which at least one phosphate group is also attached. The bases of nucleotides are planar, aromatic, heterocyclic molecules that are structural derivatives of either purine or pyrimidine (although they are not synthesised in vivo from either of these organic compounds).

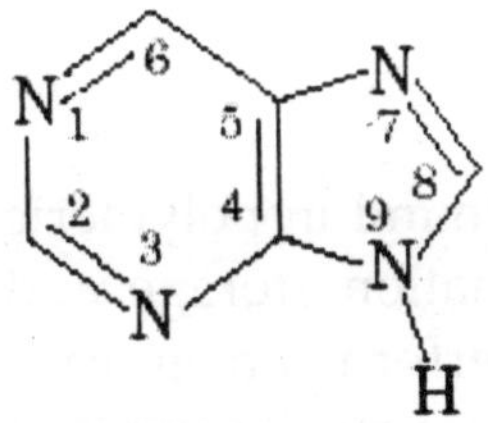

Purine

Pyrimidine

The most common purines are adenine (A) and guanine (G), and the major pyrimidines are cytosine (c), uracil (U), and thymine (T). The purines form bonds to a five-carbon sugar (a pentose) via their N9 atoms, whereas pyrimidines do so through their N1 atoms. In ribonucleotides, the pentose is ribose, while in deoxyribonucleotides (or just deoxynucleotides), the sugar is 2′-deoxyribose (i.e., the carbon at position 2′ lacks a hydroxyl group).

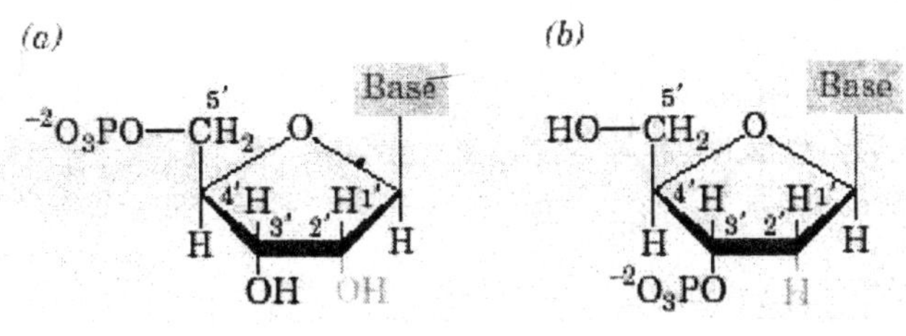

5¢-Ribonucleotide 3¢-Deoxynucleotide

Figure 7. Chemical structures of nucleotides

In a ribonucleotide or a deoxyribonucleotide, one or more phosphate groups is bonded to atom C3′ or atom C5′ of the pentose to form a 3′-nucleotide or a 5′-nucleotide, respectively (Fig. 7).When the phosphate group is absent, the compound is known as a nucleoside. A 5′-nucleotide can therefore be called a nucleoside-5′-phosphate. Nucleotides most commonly contain one to three phosphate groups at the C5′ position and are called nucleoside monophosphates, diphosphates, and triphosphates.

Ribonucleotides are found in RNA (ribonucleic acid), whereas deoxynucleotides are found in DNA (deoxyribonucleic acid). Adenine, guanine, and cytosine are found as both ribonucleotides and deoxynucleotides (accounting for six of the eight common nucleotides), but uracil is found primarily as a ribonucleotide and thymine as a deoxynucleotide. Free nucleotides, which are anionic, are usually associated with the counterion Mg^{2+} in cells.

ATP and Nucleotide Derivatives

The bulk of the nucleotides in any cell are found in polymeric forms, as either DNA or RNA, whose primary functions are information storage and transfer. However, free nucleotides and nucleotide derivatives perform an enormous variety of metabolic functions not related to the management of genetic information.Perhaps the best known nucleotide is adenosine triphosphate (ATP), a nucleotide containing adenine, ribose, and a triphosphate group.ATP is often mistakenly referred to as an energy-storage molecule, but it is more accurately termed an energy carrier or energy transfer agent. The process of photosynthesis or the breakdown of metabolic fuels such as carbohydrates and fatty acids leads to the formation of ATP from adenosine diphosphate (ADP):

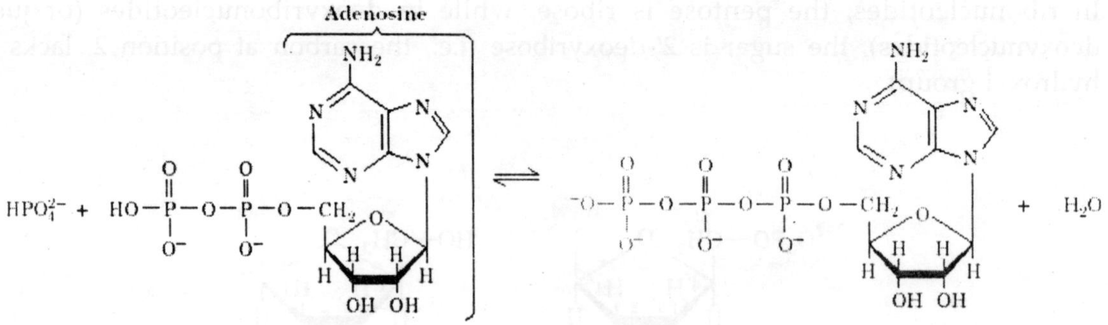

Figure 8. Adenosine diphosphate (ADP) and Adenosine triphosphate (ATP)

ATP diffuses throughout the cell to provide energy for other cellular work, such as biosynthetic reactions, ion transport, and cell movement.The chem-ical potential energy of ATP is made available when it transfers one (or two) of its phosphate groups to another molecule. This process can be represented by the reverse of the preceding reaction, namely, the hydrolysis of ATP to ADP. The degree to which ATP participates in routine cellular activities is illustrated by calculations indicating that while the concentration of cellular ATP is relatively moderate (~5 mM), the average human recycles his or her own weight of ATP each day. Nucleotide derivatives participate in a wide variety of metabolic processes. For example, starch synthesis in plants proceeds by repeated additions of glucose units donated by ADP–glucose (Fig. 8). The attached group, which may be a small molecule such as glucose or even another nucleotide, is typically linked to the nucleotide through a mono- or diphosphate group.

Nucleic Acid Structure

Nucleotides can be joined to each other to form the polymers that are familiar to us as RNA and DNA. The nucleic acids are chains of nucleotides whose phosphates bridge the 3′ and 5′ positions of neighboring ribose units. The phosphates of these polynucleotides are acidic, so at physiological pH, nucleic acids are polyanions.The linkage between individual nucleotides is known as a phosphodiester bond, so named because the phosphate is esterified to two ribose units. Each nucleotide that has been incorporated into the polynucleotide is known as a nucleotide residue.The terminal residue whose C5′ is not linked to another nucleotide is called the 5′ end, and the terminal residue whose C3′ is not linked to another nucleotide is called the 3′ end. By convention, the sequence of nucleotide residues in a nucleic acid is written, left to right, from the 5′ end to the 3′ end. The properties of a polymer such as a nucleic acid may be very different from the properties of the individual units, or monomers, before polymerisation. As the size of the polymer increases from dimer, trimer, tetramer, and so on through oligomer, physical properties such as charge and solubility may change. In addition, *a polymer of* nonidentical residues has a property that its component monomers do not have—namely, it contains information in the form of its sequence of residues.

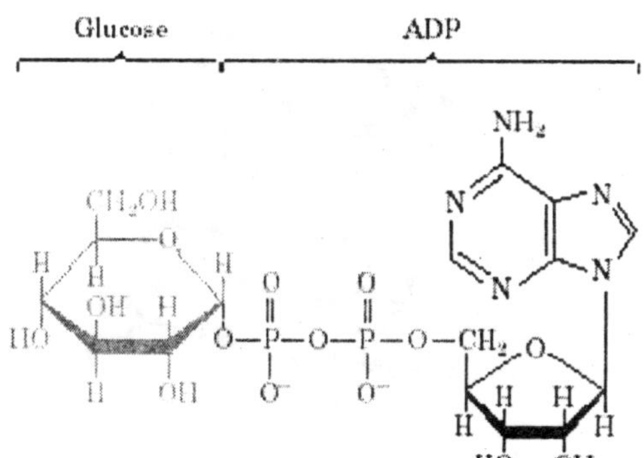

Figure 9. ADP–glucose. In this nucleotide derivative, glucose is attached to adenosine by a diphosphate group.

Although there appear to be no rules governing the nucleotide composition of typical RNA molecules, DNA has equal numbers of adenine and thymine residues (A = T) and equal numbers of guanine and cytosine residues (G = C).These relationships, known as Chargaff's rules, were discovered in the late 1940s by Erwin Chargaff, who devised the first reliable quantitative methods for the compositional analysis of DNA. DNA's base composition varies widely among different organisms. It ranges from ~25 to 75 mol % G + C in different species of bacteria. However, it is more or less constant among related species; for example, in mammals G + C ranges from 39 to 46%. The significance of Chargaff's rules was not immediately appreciated, but we now know that the structural basis for the rules derives from DNA's double-stranded nature.

Double Helix Structure

The determination of the structure of DNA by James Watson and Francis Crick in 1953 is often said to mark the birth of modern molecular biology. The Watson–Crick structure of DNA not only provided a model of what is arguably the central molecule of life, it also suggested the molecular mechanism of heredity. Watson and Crick's accomplishment, which is ranked as one of science's major intellectual achievements, was based in part on two pieces of evidence in addition to Chargaff's rules: the correct tautomeric forms of the bases and indications that DNA is a helical molecule.The purine and pyrimidine bases of nucleic acids can assume different tautomeric forms.

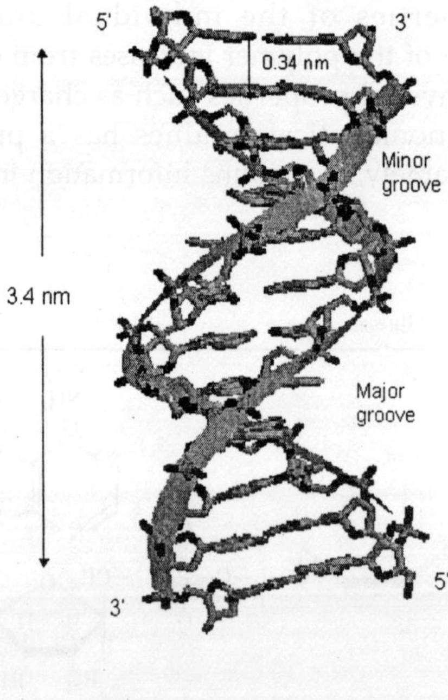

Figure 10. Double Helix Structure

X-Ray, nuclear magnetic resonance (NMR), and spectroscopic investigations have firmly established that the nucleic acid bases are overwhelmingly in the keto tautomeric forms. In 1953, however, this was not generally appreciated. Information about the dominant tautomeric forms was provided by Jerry Donohue, an office mate of Watson and Crick and an expert on the X-ray structures of small organic molecules.Evidence that DNA is a helical molecule was provided by an X-ray diffraction photograph of a DNA fiber taken by Rosalind Franklin. The appearance of the photograph enabled Crick, an X-ray crystallographer by training, to deduce (a) that DNA is a helical molecule and (b) that its planar aromatic bases form a stack that is parallel to the fiber axis.

The limited structural information, along with Chargaff's rules, provided but few clues to the structure of DNA;Watson and Crick's model sprang mostly from their imaginations and model-building studies. Once the Watson–Crick model had been published, however, its basic simplicity combined with its obvious biological relevance led to its rapid acceptance. Later investigations have confirmed the general accuracy of the Watson– Crick model, although its details have been modified. The Watson–Crick model of DNA has the following major features:

1. Two polynucleotide chains wind around a common axis to form a double helix.
2. The two strands of DNA are antiparallel (run in opposite directions), but each forms a right-handed helix.
3. The bases occupy the core of the helix and sugar–phosphate chains run along the periphery, thereby minimising the repulsions between charged phosphate groups. The surface of the double helix contains two grooves of unequal width: the major and minor grooves.
4. Each base is hydrogen bonded to a base in the opposite strand to form a planar base pair. The Watson–Crick structure can accommodate only two types of base pairs. Each adenine residue must pair with a thymine residue and vice versa, and each guanine residue must pair with a cytosine residue and vice versa.These hydrogenbonding interactions, a phenomenon known as complementary base pairing, result in the specific association of the two chains of the double helix.

The Watson–Crick structure can accommodate any sequence of bases on one polynucleotide strand if the opposite strand has the complementary base sequence. This immediately accounts for Chargaff's rules. More importantly, it suggests that each DNA strand can act as a template for the synthesis of its complementary strand and hence that hereditary information is encoded in the sequence of bases on either strand.

Most DNA molecules are extremely large, in keeping with their role as the repository of a cell's genetic information. Of course, an organism's genome, its unique DNA content, may be allocated among several chromosomes, each of which contains a separate DNA molecule. Note that many organisms are diploid; that is, they contain two equivalent sets of chromosomes, one from each parent. Their content of unique (haploid) DNA is half their total DNA. For example, humans are diploid organisms that carry 46 chromosomes per cell; the haploid number is therefore 23.

Because of their great lengths, DNA molecules are described in terms of the number of base pairs (bp) or thousands of base pairs (kilobase pairs, or kb). Naturally occurring DNAs vary in length from ~5 kb in small DNAcontaining viruses to well over 250,000 kb in the largest mammalian chromosomes. Although DNA molecules are long and relatively stiff, they are not completely rigid.We shall see later that the DNA double helix forms coils and loops when it is packaged inside the cell. Furthermore, depending on the nucleotide sequence, DNA may adopt slightly different helical conformations. Finally, in the presence of other cellular components, the DNA may bend sharply or the two strands may partially unwind.

Nucleic Acid Functions

DNA is the carrier of genetic information in all cells and in many viruses. Yet a period of over 75 years passed from the time the laws of inheritance were discovered by Gregor Mendel until the biological role of DNA was elucidated. Even now, many details of how genetic information is expressed and transmitted to future generations are still unclear.

Mendel's work with garden peas led him to postulate that an individual plant contains a pair of factors (which we now call genes), one inherited from each parent. But Mendel's theory of inheritance, reported in 1866, was almost universally ignored by his contemporaries, whose knowledge of anatomy and physiology provided no basis for its understanding. Eventually, genes were hypothesised to be part of chromosomes, and the pace of genetic research greatly accelerated.

Transmitting Genetic Information

Until the 1940s, it was generally assumed that genes were made of protein, since proteins were the only biochemical entities that, at the time, seemed complex enough to serve as agents of inheritance. Nucleic acids, which had first been isolated in 1869 by Friedrich Miescher, were believed to have monotonously repeating nucleotide sequences and were therefore unlikely candidates for transmitting genetic information.It took the efforts of Oswald Avery, Colin MacLeod, and Maclyn McCarty to demonstrate that DNA carries genetic information. Their experiments, completed in 1944, showed that DNA—not protein—extracted from a virulent (pathogenic) strain of the bacterium *Diplococcus pneumoniae* was the substance that transformed (permanently changed) a nonpathogenic strain of the organism to the virulent strain. Avery's discovery was initially greeted with skepticism, but it influenced Erwin Chargaff, whose rules led to subsequent models of the structure and function of DNA.

The double-stranded, or duplex, nature of DNA facilitates its replication. When a cell divides, each DNA strand acts as a template for the assembly of its complementary strand (Fig. 11). Consequently, every progeny cell contains a complete DNA molecule (or a set of DNA molecules in organisms whose genomes contain more than one chromosome). Each DNA molecule consists of one parental strand and one daughter strand. Daughter strands are synthesized by the stepwise polymerisation of nucleotides that specifically pair with bases on the parental strands.The mechanism of replication, while straightforward in principle, is exceedingly complex in the cell, requiring a multitude of cellular factors to proceed with fidelitv and efficiencv.

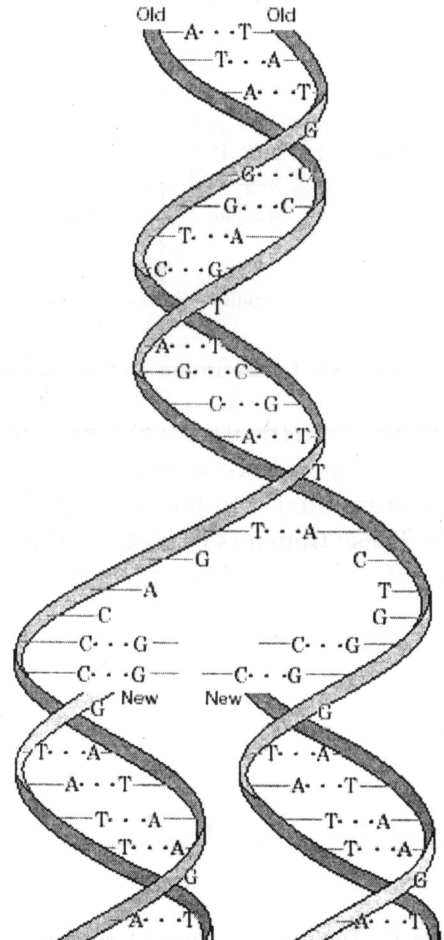

Figure 11. DNA replication

Genes Direct Protein Synthesis

The question of how sequences of nucleotides control the characteristics of organisms took some time to be answered. In experiments with the mold Neurospora crassa in the 1940s, George Beadle and Edward Tatum found that there is a specific connection between genes and enzymes, the one gene–one enzyme theory. Beadle and Tatum showed that mutant varieties of Neurospora that were generated by irradiation with X-rays required additional nutrients in order to grow. Presumably, the offspring of the radiation-damaged cells lacked the specific enzymes necessary to synthesise those nutrients.

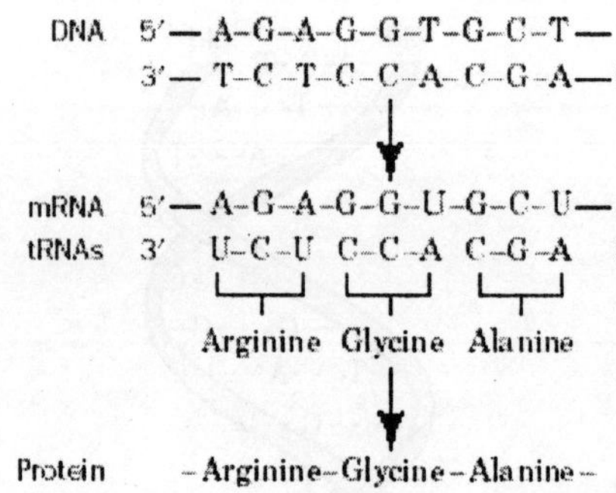

Figure 12. Transcription and translation

The link between DNA and enzymes (nearly all of which are proteins) is RNA. The DNA of a gene is transcribed to produce an RNA molecule that is complementary to the DNA. The RNA sequence is then translated into the corresponding sequence of amino acids to form a protein (Fig. 12). These transfers of biological information are summarised in the so-called *central dogma of molecular biology* formulated by Crick in 1958 (Fig. 13).

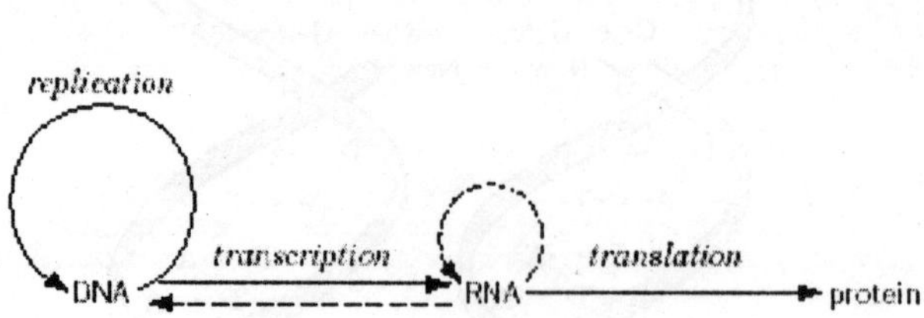

Figure 13. The central dogma of molecular biology

Just as the daughter strands of DNA are synthesised from free deoxynucleoside triphosphates that pair with bases in the parent DNA strand, RNA strands are synthesised from free ribonucleoside triphosphates that pair with the complementary bases in one

DNA strand of a gene. The RNA that corresponds to a protein-coding gene (called *messenger RNA*, or *mRNA*) makes its way to a *ribosome*, an organelle that is itself composed largely of RNA *(ribosomal RNA*, or *rRNA)*. At the ribosome, each set of three nucleotides in the mRNA pairs with three complementary nucleotides in a small RNA molecule—a *transfer RNA*, or *tRNA*.

Attached to each tRNA molecule is its corresponding amino acid. The ribosome catalyzes the joining of amino acids, which are the monomeric units of proteins. Amino acids are added to the growing protein chain according to the order in which the tRNA molecules bind to the mRNA. Since the nucleotide sequence of the mRNA in turn reflects the sequences of nucleotides in the gene, DNA directs the synthesis of proteins. It follows that alterations to the genetic material of an organism *(mutations)* may manifest themselves as proteins with altered structures and functions.

NUCLEIC ACID SEQUENCING

The ability to determine the sequence of nucleotides in nucleic acids has made it possible to deduce the amino acid sequences of their encoded proteins and, to some extent, the structures and functions of those proteins. Nucleic acid sequencing has also revealed information *about the regulation of genes*. Portions of genes that are not actually transcribed into RNA nevertheless may influence how often a gene is transcribed and translated, that is, expressed. Moreover, efforts to elucidate the sequences in hitherto unmapped regions of DNA have led to the discovery of new genes and new regulatory elements. *Once in hand, a nucleic* acid sequence can be duplicated, modified, and expressed, making it possible to study proteins that could not otherwise be obtained in useful quantities.

The overall strategy for sequencing any polymer of nonidentical units is

1. Cleave the polymer into specific fragments that are small enough to be fully sequenced.
2. Determine the sequence of residues in each fragment.
3. Determine the order of the fragments in the original polymer by repeating the preceding steps using a degradation procedure that yields a set of fragments that overlap the cleavage points in the first step.

The first efforts to sequence RNA used nonspecific enzymes to generate relatively small fragments whose nucleotide composition was then determined by partial digestion with an enzyme that selectively removed nucleotides from one end or the other. Sequencing RNA in this manner was tedious and time-consuming. Using such methods, it took Robert Holley seven years to determine the sequence of a 76-residue tRNA molecule.

After 1975, dramatic progress was made in nucleic acid sequencing technology. The advances were made possible by the discovery of enzymes that could cleave DNA at specific sites and by the development of rapid sequencing techniques for DNA. The advent of modern molecular cloning techniques also made it possible to produce sufficient quantities of specific DNA to be sequenced. These cloning techniques are necessary because most specific DNA sequences are normally present in a genome in only a single copy.

SACCHARIDES

Monosaccharides are the simplest form of carbohydrates with only one simple sugar. They essentially contain an aldehyde or ketone group in their structure. Examples of monosaccharides are the hexoses glucose, fructose, and galactose and pentoses, ribose, and deoxyribose. Disaccharides are formed when two monosaccharides, or two single simple sugars, form a bond with removal of water. Examples of disaccharides include sucrose, maltose, and lactose. Polysaccharides are polymerized monosaccharides, complex, carbohydrates. They have multiple simple sugars. Examples are starch, cellulose, and glycogen. They are generally large and often have a complex branched connectivity. Shorter polysaccharides, with 2-10 monomers, are called oligosaccharides.

Monosaccharides

Monosaccharides are the most basic unit of carbohydrates. They are the simplest form of sugar and are usually colorless, water-soluble, crystalline solids. Some monosaccharides have a sweet taste. Examples of monosaccharides include glucose (dextrose), fructose (levulose), galactose, xylose and ribose. Monosaccharides are the building blocks of disaccharides such as sucrose and polysaccharides (such as cellulose and starch). Further, each carbon atom that supports a hydroxyl group (except for the first and last) is chiral, giving rise to a number of isomeric forms all with the same chemical formula. For instance, galactose and glucose are both aldohexoses, but have different chemical and physical properties.

With few exceptions (e.g., deoxyribose), monosaccharides have the chemical formula $Cx(H_2O)y$ with the chemical structure $H(CHOH)_nC{=}O(CHOH)_mH$. If n or m is zero, it is an aldehyde and is termed an aldose, otherwise it is a ketone and is termed a ketose. Monosaccharides contain either a ketone or aldehyde functional group, and hydroxyl groups on most or all of the non-carbonyl carbon atoms.

Disaccharides

A disaccharide is the carbohydrate formed when two monosaccharides undergo a condensation reaction which involves the elimination of a small molecule, such as water, from the functional groups only. Like monosaccharides, disaccharides also dissolve in water, taste sweet and are called sugars.'Disaccharide' is one of the four chemical groupings of carbohydrates (monosaccharide, disaccharide, oligosaccharide, and polysaccharide).There are two basic types of disaccharides: reducing disaccharides, in which the monosaccharide components are bonded by hydroxyl groups; and non-reducing disaccharides, in which thc components bond through their anometric centers.

It is formed when two monosaccharides are joined together and a molecule of water is removed. For example; milk sugar (lactose) is made from glucose and galactose whereas cane sugar (sucrose) is made from glucose and fructose.The two monosaccharides are bonded via a dehydration reaction (also called a condensation reaction or dehydration synthesis) that leads to the loss of a molecule of water and formation of a glycosidic bond.

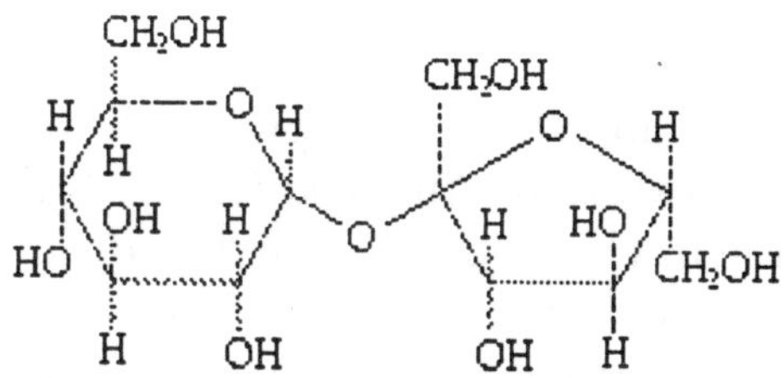

Figuer 14. Sucrose

The glycosidic bond can be formed between any hydroxyl group on the component monosaccharide. So, even if both component sugars are the same (e.g., glucose), different bond combinations (regiochemistry) and stereochemistry (alpha- or beta-) result in disaccharides that are diastereoisomers with different chemical and physical properties. Depending on the monosaccharide constituents, disaccharides are sometimes crystalline, sometimes water-soluble, and sometimes sweet-tasting and sticky-feeling.

Common disaccharides

Disaccharide	Unit 1	Unit 2
Sucrose	glucose	fructose
Lactulose	galactose	fructose
Lactose (milk sugar)	galactose	glucose
Maltose	glucose	glucose
Trehalose	glucose	glucose
Cellobiose	glucose	glucose

Maltose and cellobiose are hydrolysis products of the polysaccharides, starch and cellulose, respectively.

Polysaccharides

Polysaccharides are polymeric carbohydrate structures, formed of repeating units (either mono- or di-saccharides) joined together by glycosidic bonds. These structures are often linear, but may contain various degrees of branching. Polysaccharides are often quite heterogeneous, containing slight modifications of the repeating unit. Depending on the structure, these macromolecules can have distinct properties from their monosaccharide building blocks. They may be amorphous or even insoluble in water. When all the monosaccharides in a polysaccharide are the same type the polysaccharide is called a homopolysaccharide, but when more than one type of monosaccharide is present they are called heteropolysaccharides.Examples include storage polysaccharides such as starch and glycogen, and structural polysaccharides such as cellulose and chitin.

Starches

Starches are glucose polymers in which glucopyranose units are bonded by alpha-linkages. It is made up of a mixture of Amylose (15-20%) and Amylopectin (80-85%). Amylose consists of a linear chain of several hundred glucose molecules and Amylopectin is a branched molecule made of several thousand glucose units(every chain 24-30 glucose unit). Starches are insoluble in water. They can be digested by hydrolysis, catalyzed by enzymes called amylases, which can break the alpha-linkages (glycosidic bonds). Humans and other animals have amylases, so they can digest starches. Potato, rice, wheat, and maize are major sources of starch in the human diet. The formation of starches are the way that plants store glucose.

Glycogen

Glycogen is a polysaccharide that is found in animals and is composed of a branched chain of glucose residues. It is stored in liver and muscles.

Cellulose

The structural component of plants are formed primarily from cellulose. Wood is largely cellulose and lignin, while paper and cotton are nearly pure cellulose. Cellulose is a polymer made with repeated glucose units bonded together by beta-linkages. Humans and many other animals lack an enzyme to break the beta-linkages, so they do not digest cellulose. Certain animals can digest cellulose, because bacteria possessing the enzyme are present in their gut. The classic example is the termite.

Chitin

Chitin is one of many naturally occurring polymers. It is one of the most abundant natural materials in the world. Over time it is bio-degradable in the natural environment. Its breakdown may be catalyzed by enzymes called chitinases, secreted by microorganisms such as bacteria and fungi, and produced by some plants. Some of these microorganisms have receptors to simple sugars from the decomposition of chitin. If chitin is detected, they then produce enzymes to digest it by cleaving the glycosidic bonds in order to convert it to simple sugars and ammonia.Chemically, chitin is closely related to chitosan (a more water-soluble derivative of chitin). It is also closely related to cellulose in that it is a long unbranched chain of glucose derivatives. Both materials contribute structure and strength, protecting the organism.

Acidic polysaccharides

Acidic polysaccharides are polysaccharides that contain carboxyl groups, phosphate groups and/or sulfuric ester groups.

Bacterial polysaccharides

Bacterial polysaccharides represent a diverse range of macromolecules that include peptidoglycan, lipopolysaccharides, capsules and exopolysaccharides; compounds whose functions range from structural cell-wall components (eg. peptidoglycan), and important virulence factors (eg Poly-N-acetylglucosamine in S. aureus), to permitting the bacterium to survive in harsh environments (eg Pseudomonas aeruginosa in the human lung). Polysaccharide biosynthesis is a tightly regulated, energy intensive process and understanding the subtle interplay between the regulation and energy conservation, polymer modification and synthesis, and the external ecological functions is a huge area of research. The potential benefits are enormous and should enable for example the development of novel antibacterial strategies (eg new antibiotics and vaccines) and the commercial exploitation to develop novel applications.

Pathogenic bacteria commonly produce a thick, mucous-like, layer of polysaccharide. This "capsule" cloaks antigenic proteins on the bacterial surface that would otherwise provoke an immune response and thereby lead to the destruction of the bacteria. Capsular polysaccharides are water soluble, commonly acidic, and have molecular weights on the order of 100-1000 kDa. They are linear and consist of regularly repeating subunits of one to six monosaccharides. There is enormous structural diversity; nearly two hundred different polysaccharides are produced by E. coli alone. Mixtures of capsular polysaccharides, either conjugated or native are used as vaccines.

Bacteria and many other microbes, including fungi and algae, often secrete polysaccharides as an evolutionary adaptation to help them adhere to surfaces and to prevent them from drying out. Humans have developed some of these polysaccharides into useful products, including xanthan gum, dextran, gellan gum, and pullulan.

Cell-surface polysaccharides play diverse roles in bacterial ecology and physiology. They serve as a barrier between the cell wall and the environment, mediate host-pathogen interactions, and form structural components of biofilms. These polysaccharides are synthesized from nucleotide-activated precursors (called nucleotide sugars) and, in most cases, all the enzymes necessary for biosynthesis, assembly and transport of the completed polymer are encoded by genes organized in dedicated clusters within the genome of the organism. Lipopolysaccharide is one of the most important cell-surface polysaccharides, as it plays a key structural role in outer membrane integrity, as well as being an important mediator of host-pathogen interactions.

The enzymes that make the A-band (homopolymeric) and B-band (heteropolymeric) O-antigens have been identified and the metabolic pathways defined. The exopolysaccharide alginate is a linear copolymer of ß-1,4-linked D-mannuronic acid and L-guluronic acid residues, and is responsible for the mucoid phenotype of late-stage cystic fibrosis disease. The pel and psl loci are two recently discovered gene clusters that also encode exopolysaccharides found to be important for biofilm formation. Rhamnolipid is a biosurfactant whose production is tightly regulated at the transcriptional level, but the precise role that it plays in disease is not well understood at present. Protein glycosylation, particularly of pilin and flagellin, is a recent focus of research by several groups and it has been shown to be important for adhesion and invasion during bacterial infection.

LIGNIN

Lignin or lignen is a complex chemical compound most commonly derived from wood, and an integral part of the secondary cell walls of plants and some algae. The term was introduced in 1819 by de Candolle and is derived from the Latin word lignum, meaning wood. It is one of the most abundant organic polymers on Earth, exceeded only by cellulose, employing 30% of non-fossil organic carbon and constituting from a quarter to a third of the dry mass of wood. As a biopolymer, lignin is unusual because of its heterogeneity and lack of a defined primary structure. Its most commonly noted function is the support through strengthening of wood (xylem cells) in trees.

Biological Function

Lignin fills the spaces in the cell wall between cellulose, hemicellulose, and pectin components, especially in tracheids, sclereids and xylem. It is covalently linked to hemicellulose and thereby crosslinks different plant polysaccharides, conferring mechanical strength to the cell wall and by extension the plant as a whole. It is particularly abundant in compression wood but scarce in tension wood.

Lignin plays a crucial part in conducting water in plant stems. The polysaccharide components of plant cell walls are highly hydrophilic and thus permeable to water, whereas lignin is more hydrophobic. The crosslinking of polysaccharides by lignin is an obstacle for water absorption to the cell wall. Thus, lignin makes it possible for the plant's vascular tissue to conduct water efficiently. Lignin is present in all vascular plants, but not in bryophytes, supporting the idea that the original function of lignin was restricted to water transport. However, it is present in red algae, which seems to suggest that the common ancestor of plants and red algae also synthesised lignin. This would suggest that its original function was structural; it plays this role in the red alga Calliarthron, where it supports joints between calcified segments.

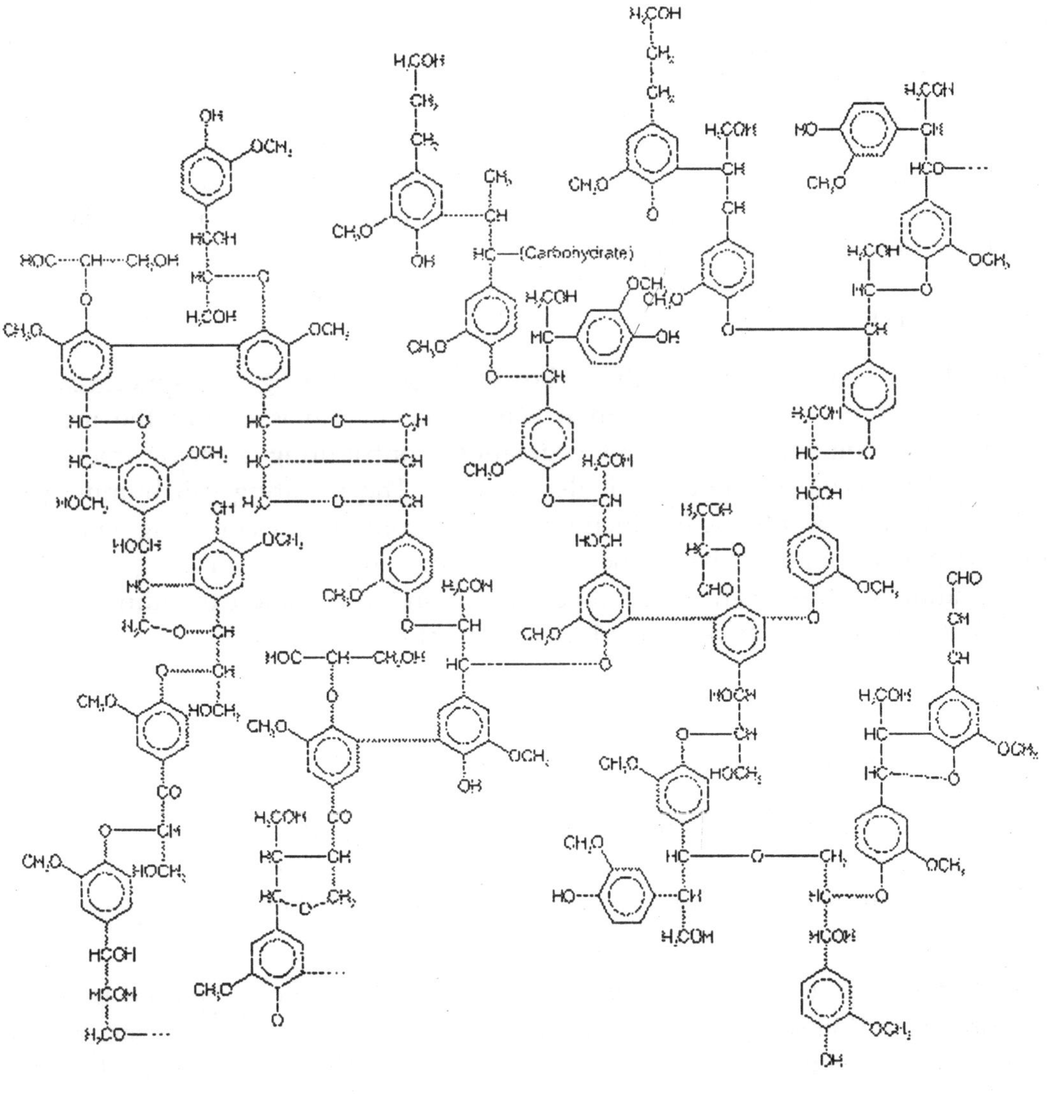

Figuer 15. An example of a possible lignin structure

Lignin is indigestible by animal enzymes, but some fungi and bacteria are able to secrete ligninases which can biodegrade the polymer. The details of the biodegradation are not well understood. The pathway depends on the type of wood decay - in fungi either brown rot, soft rot or white rot. The enzymes involved may employ free radicals for depolymerization reactions. Well understood lignolytic enzymes are manganese peroxidase, lignin peroxidase and cellobiose dehydrogenase. Furthermore, because of its cross-linking with the other cell wall components, it minimizes the accessibility of cellulose and hemicellulose to microbial enzymes. Hence, lignin is generally associated with reduced digestibility of the overall plant biomass, which helps defend against pathogens and pests.

Lignin peroxidase (also "ligninase", EC number 1.14.99) is a hemoprotein from the white-rot fungus Phanerochaete chrysosporium with a variety of lignin-degrading reactions, all dependent on hydrogen peroxide to incorporate molecular oxygen into reaction products. There are also several other microbial enzymes that are believed to be involved in lignin biodegradation, such as manganese peroxidase, laccase and Cellobiose dehydrogenase (acceptor).

Ecological Function

Lignin plays a significant role in the carbon cycle, sequestering atmospheric carbon into the living tissues of woody perennial vegetation. Lignin is one of the most slowly decomposing components of dead vegetation, contributing a major fraction of the material that becomes humus as it decomposes. The resulting soil humus generally increases the photosynthetic productivity of plant communities growing on a site as the site transitions from disturbed mineral soil through the stages of ecological succession, by providing increased cation exchange capacity in the soil and expanding the capacity of moisture retention between flood and drought conditions.

Economic Significance

Highly lignified wood is durable and therefore a good raw material for many applications. It is also an excellent fuel, since lignin yields more energy when burned than cellulose. Mechanical, or high yield pulp used to make newsprint contains most of the lignin originally present in the wood. This lignin is responsible for newsprint yellowing with age. Lignin must be removed from the pulp before high quality bleached paper can be manufactured from it.

The first investigations into commercial use of lignin were reported by Marathon Corporation in Rothschild, Wisconsin (USA), starting in 1927. The first class of products which showed promise were leather tanning agents. The lignin chemical business of Marathon was operated for many years as Marathon Chemicals. It is now known as LignoTech USA, Inc., and is owned by the Norwegian company, Borregaard, itself a subsidiary of the Norwegian conglomerate Orkla AS.

Lignin removed via the kraft process (sulfate pulping) is usually burned for its fuel value, providing more than enough energy to run the mill and its associated processes.

More recently, lignin extracted from shrubby willow has been successfully used to produce expanded polyurethane foam.

In 1998, a German company, Tecnaro, developed a process for turning lignin into a substance, called Arboform, which behaves identically to plastic for injection molding. Therefore, it can be used in place of plastic for several applications. When the item is discarded, it can be burned just like wood.

Lignin is a cross-linked racemic macromolecule with molecular masses in excess of 10,000 μ. It is relatively hydrophobic and aromatic in nature. The degree of polymerisation in nature is difficult to measure, since it is fragmented during extraction and the molecule consists of various types of substructures which appear to repeat in a haphazard manner. Different types of lignin have been described depending on the means of isolation.

There are three monolignol monomers, methoxylated to various degrees: p-coumaryl alcohol, coniferyl alcohol, and sinapyl alcohol. These lignols are incorporated into lignin in the form of the phenylpropanoids p-hydroxyphenyl (H), guaiacyl (G), and syringal (S) respectively. Gymnosperms have a lignin that consists almost entirely of G with small quantities of H. That of dicotyledonous angiosperms is more often than not a mixture of G and S (with very little H), and monocotyledonous lignin is a mixture of all three. Many grasses have mostly G, while some palms have mainly S.[citation needed] All lignins contain small amounts of incomplete or modified monolignols, and other monomers are prominent in non-woody plants.

Biosynthesis

Lignin biosynthesis begins in the cytosol with the synthesis of glycosylated monolignols from the amino acid phenylalanine. These first reactions are shared with the phenylpropanoid pathway. The attached glucose renders them water soluble and less toxic. Once transported through the cell membrane to the apoplast, the glucose is removed and the polymerisation commences.[citation needed] Much about its anabolism is not understood even after more than a century of study.

The polymerisation step, that is a radical-radical coupling, is catalysed by oxidative enzymes. Both peroxidase and laccase enzymes are present in the plant cell walls, and it is not known whether one or both of these groups participates in the polymerisation. Low molecular weight oxidants might also be involved. The oxidative enzyme catalyses the formation of monolignol radicals. These radicals are often said to undergo uncatalyzed coupling to form the lignin polymer, but this hypothesis has been recently challenged. The alternative theory that involves an unspecified biological control is however not widely accepted.Pyrolysis of lignin during the combustion of wood or charcoal production yields a range of products, of which the most characteristic ones are methoxy phenols. Of those, the most important are guaiacol and syringol and their derivatives; their presence can be used to trace a smoke source to a wood fire. In cooking, lignin in the form of hardwood is an important source of these two chemicals which impart the characteristic aroma and taste to smoked foods.

LIPIDS

Lipids are a broad group of naturally-occurring molecules which includes fats, waxes, sterols, fat-soluble vitamins (such as vitamins A, D, E and K), monoglycerides, diglycerides, phospholipids, and others. The main biological functions of lipids include energy storage, as structural components of cell membranes, and as important signaling molecules.

Lipids may be broadly defined as hydrophobic or amphiphilic small molecules; the amphiphilic nature of some lipids allows them to form structures such as vesicles, liposomes, or membranes in an aqueous environment. Biological lipids originate entirely or in part from two distinct types of biochemical subunits or "building blocks": ketoacyl and isoprene groups. Using this approach, lipids may be divided into eight categories: fatty acyls, glycerolipids, glycerophospholipids, sphingolipids, saccharolipids and polyketides (derived from condensation of ketoacyl subunits); and sterol lipids and prenol lipids (derived from condensation of isoprene subunits).

Although the term lipid is sometimes used as a synonym for fats, fats are a subgroup of lipids called triglycerides. Lipids also encompass molecules such as fatty acids and their derivatives (including tri-, di-, and monoglycerides and phospholipids), as well as other sterol-containing metabolites such as cholesterol. Although humans and other mammals use various biosynthetic pathways to both break down and synthesize lipids, some essential lipids cannot be made this way, and must be obtained from the diet.

Categories of Lipids

Fatty acyls

Fatty acyls, a generic term for describing fatty acids, their conjugates and derivatives, are a diverse group of molecules synthesized by chain-elongation of an acetyl-CoA primer with malonyl-CoA or methylmalonyl-CoA groups in a process called fatty acid synthesis. They are made of a hydrocarbon chain that terminates with a carboxylic acid group; this arrangement confers the molecule with a polar, hydrophilic end, and a nonpolar, hydrophobic end that is insoluble in water. The fatty acid structure is one of the most fundamental categories of biological lipids, and is commonly used as a building block of more structurally complex lipids. The carbon chain, typically between four to

24 carbons long, may be saturated or unsaturated, and may be attached to functional groups containing oxygen, halogens, nitrogen and sulfur. Where a double bond exists, there is the possibility of either a cis or trans geometric isomerism, which significantly affects the molecule's molecular configuration. Cis-double bounds cause the fatty acid chain to bend, an effect that is more pronounced the more double bonds there are in a chain. This in turn plays an important role in the structure and function of cell membranes. Most naturally occurring fatty acids are of the cis configuration, although the trans form does exist in some natural and partially hydrogenated fats and oils.

Examples of biologically important fatty acids are the eicosanoids, derived primarily from arachidonic acid and eicosapentaenoic acid, which include prostaglandins, leukotrienes, and thromboxanes. Other major lipid classes in the fatty acid category are the fatty esters and fatty amides. Fatty esters include important biochemical intermediates such as wax esters, fatty acid thioester coenzyme A derivatives, fatty acid thioester ACP derivatives and fatty acid carnitines. The fatty amides include N-acyl ethanolamines, such as the cannabinoid neurotransmitter anandamide.

Glycerolipids

Glycerolipids are composed mainly of mono-, di- and tri-substituted glycerols, the most well-known being the fatty acid esters of glycerol (triacylglycerols), also known as triglycerides. In these compounds, the three hydroxyl groups of glycerol are each esterified, usually by different fatty acids. Because they function as a food store, these lipids comprise the bulk of storage fat in animal tissues. The hydrolysis of the ester bonds of triacylglycerols and the release of glycerol and fatty acids from adipose tissue is called fat mobilization.

Additional subclasses of glycerolipids are represented by glycosylglycerols, which are characterized by the presence of one or more sugar residues attached to glycerol via a glycosidic linkage. Examples of structures in this category are the digalactosyldiacylglycerols found in plant membranes and seminolipid from mammalian sperm cells.

Glycerophospholipids

Glycerophospholipids, also referred to as phospholipids, are ubiquitous in nature and are key components of the lipid bilayer of cells, as well as being involved in metabolism and cell signaling. Neural tissue (including the brain) contains relatively high amounts of glycerophospholipids, and alterations in their composition has been implicated in various neurological disorders. Glycerophospholipids may be subdivided into distinct classes, based on the nature of the polar headgroup at the sn-3 position of the glycerol backbone in eukaryotes and eubacteria, or the sn-1 position in the case of archaebacteria.

Examples of glycerophospholipids found in biological membranes are phosphatidylcholine (also known as PC, GPCho or lecithin), phosphatidylethanolamine (PE or GPEtn) and phosphatidylserine (PS or GPSer). In addition to serving as a primary component of cellular membranes and binding sites for intra- and intercellular proteins, some glycerophospholipids in eukaryotic cells, such as phosphatidylinositols and phosphatidic acids are either precursors of, or are themselves, membrane-derived second messengers. Typically, one or both of these hydroxyl groups are acylated with long-chain fatty acids, but there are also alkyl-linked and 1Z-alkenyl-linked (plasmalogen) glycerophospholipids, as well as dialkylether variants in archaebacteria.

Sphingolipids

Sphingolipids are a complex family of compounds that share a common structural feature, a sphingoid base backbone that is synthesized de novo from the amino acid serine and a long-chain fatty acyl CoA, then converted into ceramides, phosphosphingolipids, glycosphingolipids and other compounds. The major sphingoid base of mammals is commonly referred to as sphingosine. Ceramides (N-acyl-sphingoid bases) are a major subclass of sphingoid base derivatives with an amide-linked fatty acid. The fatty acids are typically saturated or mono-unsaturated with chain lengths from 16 to 26 carbon atoms.

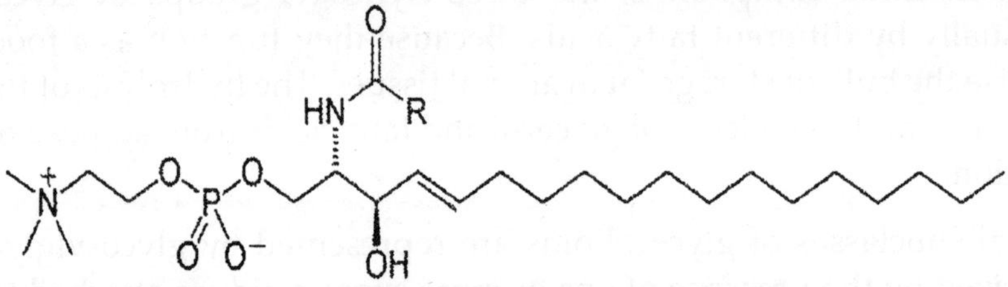

Figuer 16. Sphingomyelin

The major phosphosphingolipids of mammals are sphingomyelins (ceramide phosphocholines), whereas insects contain mainly ceramide phosphoethanolamines and fungi have phytoceramide phosphoinositols and mannose-containing headgroups. The glycosphingolipids are a diverse family of molecules composed of one or more sugar residues linked via a glycosidic bond to the sphingoid base. Examples of these are the simple and complex glycosphingolipids such as cerebrosides and gangliosides.

Sterol lipids

Sterol lipids, such as cholesterol and its derivatives, are an important component of membrane lipids, along with the glycerophospholipids and sphingomyelins. The steroids, all derived from the same fused four-ring core structure, have different biological roles as hormones and signaling molecules. The eighteen-carbon (C18) steroids include the estrogen family whereas the C19 steroids comprise the androgens such as testosterone and androsterone. The C21 subclass includes the progestogens as well as the glucocorticoids and mineralocorticoids. The secosteroids, comprising various forms of vitamin D, are characterized by cleavage of the B ring of the core structure. Other examples of sterols are the bile acids and their conjugates, which in mammals are oxidized derivatives of cholesterol and are synthesized in the liver. The plant equivalents are the phytosterols, such as ß-sitosterol, stigmasterol, and brassicasterol; the latter compound is also used as a biomarker for algal growth. The predominant sterol in fungal cell membranes is ergosterol.

Prenol lipids

Prenol lipids are synthesized from the 5-carbon precursors isopentenyl diphosphate and dimethylallyl diphosphate that are produced mainly via the mevalonic acid (MVA) pathway. The simple isoprenoids (linear alcohols, diphosphates, etc.) are formed by the successive addition of C5 units, and are classified according to number of these terpene units. Structures containing greater than 40 carbons are known as polyterpenes. Carotenoids are important simple isoprenoids that function as antioxidants and as precursors of vitamin A. Another biologically important class of molecules is exemplified by the quinones and hydroquinones, which contain an isoprenoid tail attached to a quinonoid core of non-isoprenoid origin. Vitamin E and vitamin K, as well as the ubiquinones, are examples of this class. Prokaryotes synthesize polyprenols (called bactoprenols) in which the terminal isoprenoid unit attached to oxygen remains unsaturated, whereas in animal polyprenols (dolichols) the terminal isoprenoid is reduced.

Saccharolipids

Glucosamine residues in blue, Kdo residues in red, acyl chains in black and phosphate groups in green.Saccharolipids describe compounds in which fatty acids are linked directly to a sugar backbone, forming structures that are compatible with membrane bilayers. In the saccharolipids, a monosaccharide substitutes for the glycerol backbone present in glycerolipids and glycerophospholipids. The most familiar saccharolipids are the acylated glucosamine precursors of the Lipid A component of the lipopolysaccharides in Gram-negative bacteria. Typical lipid A molecules are disaccharides of glucosamine, which are derivatized with as many as seven fatty-acyl chains. The minimal lipopolysaccharide required for growth in E. coli is Kdo2-Lipid A, a hexa-acylated disaccharide of glucosamine that is glycosylated with two 3-deoxy-D-manno-octulosonic acid (Kdo) residues.

Polyketides

Polyketides are synthesized by polymerization of acetyl and propionyl subunits by classic enzymes as well as iterative and multimodular enzymes that share mechanistic features with the fatty acid synthases. They comprise a large number of secondary metabolites and natural products from animal, plant, bacterial, fungal and marine sources, and have great structural diversity. Many polyketides are cyclic molecules whose backbones are often further modified by glycosylation, methylation, hydroxylation, oxidation, and/or other processes. Many commonly used anti-microbial, anti-parasitic, and anti-cancer agents are polyketides or polyketide derivatives, such as erythromycins, tetracyclines, avermectins, and antitumor epothilones.

Biological Functions

Membranes

Eukaryotic cells are compartmentalized into membrane-bound organelles which carry out different biological functions. The glycerophospholipids are the main structural component of biological membranes, such as the cellular plasma membrane and the intracellular membranes of organelles; in animal cells the plasma membrane physically separates the intracellular components from the extracellular environment. The glycerophospholipids are amphipathic molecules (containing both hydrophobic and hydrophilic regions) that contain a glycerol core linked to two fatty acid-derived "tails" by ester linkages and to one "head" group by a phosphate ester linkage. While glycerophospholipids are the major component of biological membranes, other non-glyceride lipid components such as sphingomyelin and sterols (mainly cholesterol in animal cell membranes) are also found in biological membranes. In plants and algae, the galactosyldiacylglycerols, and sulfoquinovosyldiacylglycerol, which lack a phosphate group, are important components of membranes of chloroplasts and related organelles and are the most abundant lipids in photosynthetic tissues, including those of higher plants, algae and certain bacteria. Bilayers have been found to exhibit high levels of birefringence which can be used to probe the degree of order (or disruption) within the bilayer using techniques such as dual polarisation interferometry.

A biological membrane is a form of lipid bilayer. The formation of lipid bilayers is an energetically-preferred process when the glycerophospholipids described above are in an aqueous environment. In an aqueous system, the polar heads of lipids align towards the polar, aqueous environment, while the hydrophobic tails minimize their contact with water and tend to cluster together, forming a vesicle; depending on the concentration of the lipid, this biophysical interaction may result in the formation of micelles, liposomes, or lipid bilayers. Other aggregations are also observed and form part of the polymorphism of amphiphile (lipid) behavior.

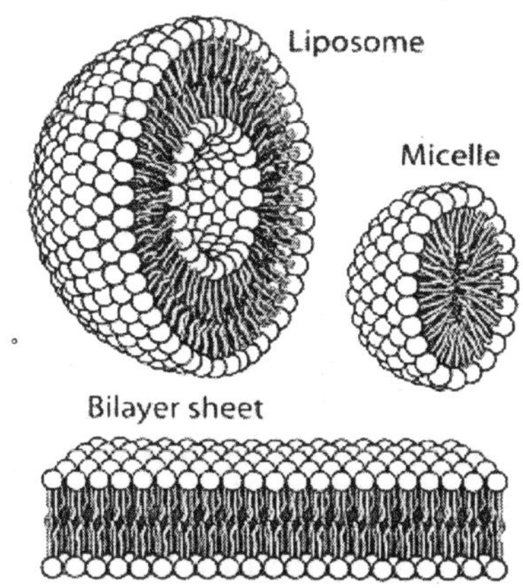

Figuer 17. Self-organization of phospholipids: a spherical liposome, a micelle and a lipid bilayer.

Phase behavior is an area of study within biophysics and is the subject of current academic research. Micelles and bilayers form in the polar medium by a process known as the hydrophobic effect. When dissolving a lipophilic or amphiphilic substance in a polar environment, the polar molecules (i.e., water in an aqueous solution) become more ordered around the dissolved lipophilic substance, since the polar molecules cannot form hydrogen bonds to the lipophilic areas of the amphiphile. So in an aqueous environment, the water molecules form an ordered "clathrate" cage around the dissolved lipophilic molecule.

Energy storage

Triacylglycerols, stored in adipose tissue, are a major form of energy storage in animals. The adipocyte, or fat cell, is designed for continuous synthesis and breakdown of triacylglycerols, with breakdown controlled mainly by the activation of hormone-sensitive enzyme lipase. The complete oxidation of fatty acids provides high caloric content, about 9 kcal/g, compared with 4 kcal/g for the breakdown of carbohydrates and proteins. Migratory birds that must fly long distances without eating use stored energy of triacylglycerols to fuel their flights.

Signaling

In recent years, evidence has emerged showing that lipid signaling is a vital part of the cell signaling. Lipid signaling may occur via activation of G protein-coupled or nuclear receptors, and members of several different lipid categories have been identified as signaling molecules and cellular messengers. These include sphingosine-1-phosphate, a sphingolipid derived from ceramide that is a potent messenger molecule involved in regulating calcium mobilization, cell growth, and apoptosis; diacylglycerol (DAG) and the phosphatidylinositol phosphates (PIPs), involved in calcium-mediated activation of protein kinase C; the prostaglandins, which are one type of fatty-acid derived eicosanoid involved in inflammation and immunity; the steroid hormones such as estrogen, testosterone and cortisol, which modulate a host of functions such as reproduction, metabolism and blood pressure; and the oxysterols such as 25-hydroxy-cholesterol that are liver X receptor agonists.

The "fat-soluble" vitamins (A, D, E and K) - which are isoprene-based lipids - are essential nutrients stored in the liver and fatty tissues, with a diverse range of functions. Acyl-carnitines are involved in the transport and metabolism of fatty acids in and out of mitochondria, where they undergo beta oxidation. Polyprenols and their phosphorylated derivatives also play important transport roles, in this case the transport of oligosaccharides across membranes. Polyprenol phosphate sugars and polyprenol diphosphate sugars function in extra-cytoplasmic glycosylation reactions, in extracellular polysaccharide biosynthesis (for instance, peptidoglycan polymerization in bacteria), and in eukaryotic protein N-glycosylation. Cardiolipins are a subclass of glycerophospholipids containing four acyl chains and three glycerol groups that are particularly abundant in the inner mitochondrial membrane. They are believed to activate enzymes involved with oxidative phosphorylation.

Metabolism

The major dietary lipids for humans and other animals are animal and plant triglycerides, sterols, and membrane phospholipids. The process of lipid metabolism synthesizes and degrades the lipid stores and produces the structural and functional lipids characteristic of individual tissues.

Biosynthesis

In animals, when there is an oversupply of dietary carbohydrate, the excess carbohydrate is converted to triacylglycerol. This involves the synthesis of fatty acids from acetyl-CoA and the esterification of fatty acids in the production of triacylglycerol, a process called lipogenesis. Fatty acids are made by fatty acid synthases that polymerize and then reduce acetyl-CoA units. The acyl chains in the fatty acids are extended by a cycle of reactions that add the acetyl group, reduce it to an alcohol, dehydrate it to an alkene group and then reduce it again to an alkane group. The enzymes of fatty acid biosynthesis are divided into two groups, in animals and fungi all these fatty acid synthase reactions are carried out by a single multifunctional protein, while in plant plastids and bacteria separate enzymes perform each step in the pathway. The fatty acids may be subsequently converted to triacylglycerols that are packaged in lipoproteins and secreted from the liver.

The synthesis of unsaturated fatty acids involves a desaturation reaction, whereby a double bond is introduced into the fatty acyl chain. For example, in humans, the desaturation of stearic acid by stearoyl-CoA desaturase-1 produces oleic acid. The doubly-unsaturated fatty acid linoleic acid as well as the triply-unsaturated linolenic acid cannot be synthesized in mammalian tissues, and are therefore essential fatty acids and must be obtained from the diet.

Triacylglycerol synthesis takes place in the endoplasmic reticulum by metabolic pathways in which acyl groups in fatty acyl-CoAs are transferred to the hydroxyl groups of glycerol-3-phosphate and diacylglycerol.

Terpenes and isoprenoids, including the carotenoids, are made by the assembly and modification of isoprene units donated from the reactive precursors isopentenyl pyrophosphate and dimethylallyl pyrophosphate. These precursors can be made in different ways. In animals and archaea, the mevalonate pathway produces these compounds from acetyl-CoA, while in plants and bacteria the non-mevalonate pathway uses pyruvate and glyceraldehyde 3-phosphate as substrates. One important reaction that uses these activated isoprene donors is steroid biosynthesis. Here, the isoprene units are joined together to make squalene and then folded up and formed into a set of rings to make lanosterol. Lanosterol can then be converted into other steroids such as cholesterol and ergosterol.

Degradation

Beta oxidation is the metabolic process by which fatty acids are broken down in the mitochondria and/or in peroxisomes to generate acetyl-CoA. For the most part, fatty acids are oxidized by a mechanism that is similar to, but not identical with, a reversal of the process of fatty acid synthesis. That is, two-carbon fragments are removed sequentially from the carboxyl end of the acid after steps of dehydrogenation, hydration, and oxidation to form a beta-keto acid, which is split by thiolysis. The acetyl-CoA is then ultimately converted into ATP, CO_2, and H_2O using the citric acid cycle and the electron transport chain. The energy yield of the complete oxidation of the fatty acid palmitate is 106 ATP. Unsaturated and odd-chain fatty acids require additional enzymatic steps for degradation.

Nutrition and health

Most of the lipid found in food is in the form of triacylglycerols, cholesterol and phospholipids. A minimum amount of dietary fat is necessary to facilitate absorption of fat-soluble vitamins (A, D, E and K) and carotenoids. Humans and other mammals have a dietary requirement for certain essential fatty acids, such as linoleic acid (an omega-6 fatty acid) and alpha-linolenic acid (an omega-3 fatty acid) because they cannot be synthesized from simple precursors in the diet. Both of these fatty acids are 18-carbon polyunsaturated fatty acids differing in the number and position of the double bonds. Most vegetable oils are rich in linoleic acid (safflower, sunflower, and corn oils). Alpha-linolenic acid is found in the green leaves of plants, and in selected seeds, nuts and legumes (particularly flax, rapeseed, walnut and soy). Fish oils are particularly rich in the longer-chain omega-3 fatty acids eicosapentaenoic acid (EPA) and docosahexaenoic acid (DHA). A large number of studies have shown positive health benefits associated with consumption of omega-3 fatty acids on infant development, cancer, cardiovascular diseases, and various mental illnesses, such as depression, attention-deficit hyperactivity disorder, and dementia. In contrast, it is now well-established that consumption of trans fats, such as those present in partially hydrogenated vegetable oils, are a risk factor for cardiovascular disease.

A few studies have suggested that total diary fat intake is linked to an increased risk of obesity and diabetes. However, a number of very large studies, including the Women's Health Initiative Dietary Modification Trial, an eight year study of 49,000 women, the Nurses' Health Study and the Health Professionals Follow-up Study, revealed no such links. None of these studies suggested any connection between percentage of calories from fat and risk of cancer, heart disease or weight gain. The Nutrition Source, a website maintained by the Department of Nutrition at the Harvard School of Public Health, summarizes the current evidence on the impact of dietary fat: "Detailed research-much of it done at Harvard-shows that the total amount of fat in the diet isn't really linked with weight or disease."

REFERENCES

Ausubel, F.M., Brent, R., Kingston, R.E., Moore, D.D., Seidman, J.G., Smith, J.A., and Struhl, K., *Short Protocols in Molecular Biology*, (4th ed.),Wiley, 1999.

Boyer, Rodney, *Concepts in Biochemistry*, Pacific Grove, CA: Brooks/Cole Publishing Company, 1999.

Jorde, L.B., J.C. Carey, M.J. Bamshad, and R.L. White, *Medical Genetics*, 2nd ed. St. Louis: Mosby-Year Book, Inc., 2000.

Lodish, H., et. al., *Molecular Cell Biology*, 4th ed. New York: W. H. Freeman & Co., 2000.

Nelson, David L., David L. Nelson, and Michael M. Cox, *Lehninger Principles of Biochemistry*, 3rd edition. Worth Publishing, 2000.

Mouritsen O. G., *Physics in Living Matter*, eds. D. Baeriswyl, M. Droz, A. Malaspinas, and P. Martinoli, Springer, Berlin, 1987.

Sakmann E., *Biophysics*, eds. W. Hoppe, W. Lohmann, H. Markl, and H. Ziegler, Springer, Berlin, 1983.

Stryer L., *Biochemistry*, 4th ed., W. H. Freeman, New York, 1995.

Watts A. and J.J.H.H.M, De Pont, *Progress in Protein-Lipie Interactions*, 2 Vols., Eds. Elsevier, New York, 1985-1986.

3

DNA Biochemistry

The human body consists of trillions of cells. Almost all contain an entire genome—the complete set of inherited genetic information encoded in our DNA. When humans reproduce, the parents' sperm and egg DNA combine to contribute a genome's worth of genetic information to the fertilized embryo. That same information is in each of the cells that eventually make up an organism.

Some segments of DNA, called genes or "coding" DNA, contain the chemical recipe that determines particular traits; genetics is the study of the inheritance and function of these genes. Scientists now estimate that humans have about 30,000 genes, located along threadlike, tightly coiled strands of DNA called chromosomes. Genes, however, are only about three percent of human DNA; the rest is "noncoding" DNA. Within these noncoding regions of the genome is the information that determines when and where genes are active—for example, in which cell types and at what stages in the life of an organism. Genomics is the study of the entire set of DNA sequences—both coding and noncoding DNA.

Over the past decade, the decoding of the genomes of human beings and other important organisms has sparked an extraordinary biological revolution. The information and technology of genomics is transforming our understanding of human evolution, the mechanisms of disease, the relationship between heredity and environment, and our ancient connection with all forms of life. In the next few years we will see many exciting discoveries leading to a better understanding of the complexity of life, as well as new drugs, vaccines, and diagnostics and less expensive, more efficient, and safer ways to restore the environment.

Some human diseases and defects are directly or indirectly caused by genetic abnormalities. Sickle cell anemia, for example, is caused by a change in just one nucleotide out of six billion. Specific genes have been associated with breast cancer, deafness, and blindness. Some illnesses are caused by complex, interacting environmental and genetic factors and cannot be explained by classical inheritance patterns. Genome studies help medical researchers understand the molecular details of these diseases so they can pursue innovative drug treatments and more quickly identify high-risk individuals who could benefit from early medical intervention. And the analysis of the genomes of disease-causing microbes, viruses, and insects, such as the human malaria parasite and its carrier, the Anopheles mosquito, are helping in the development of new prevention and treatment strategies.

Fungi and other plant pathogens cause billions of dollars in damage every year to agricultural crops, plants, and trees. Sequencing their genomes is helping botanists and foresters find effective treatments. Better understanding of plant genetics is also improving crop yields and enhancing the nutritional value of food.

Microbes—nature's simplest and most abundant organisms—can thrive under extreme conditions of heat, cold, pressure, and even radiation. By studying their genomes, scientists hope to find ways to use bacteria and other microorganisms to solve a variety of environmental problems, develop new energy sources, and improve industrial processes. Some microbes can help clean up hazardous waste sites by absorbing, transforming, or breaking down contaminants—a technique called bioremediation. Others can help combat global warming by absorbing, or sequestering, carbon from the atmosphere. And microbes can convert a wide range of organic and inorganic materials into renewable energy

DNA: LIFE'S CODE

The double-stranded DNA (deoxyribonucleic acid) molecule contains the four basic chemical units of life's code: the nucleotide bases adenine (A), guanine (G), cytosine (C), and thymine (T). These combine into the base pairs AT, TA, GC, and CG. The paired bases form the "rungs" of a structure that looks like a twisted rope ladder—the famous double helix. Sugar and phosphate molecules form the outer edges.

The DNA Sequence in every human is 99.9 percent identical to that of every other human. The slight variations in our genomes are called single nucleotide polymorphisms, or SNPs. Scientists estimate that there are about 1.4 million locations on the genome where SNPs occur in humans. It is these small variations that contribute to individual differences. SNPs and other mutations can be caused by copying errors as DNA is reproduced, or triggered by radiation, viruses, or toxic substances in the environment. Many SNPs have no effect on cell function, but others can cause or predispose a person to disease or influence response to a drug.

Following the DNA recipe, our cells manufacture the proteins that are responsible for the structure and functioning of our bodies. Proteins are involved in many of the body's life processes, including growth, repair, digestion, and aging. Many proteins are enzymes that can trigger or accelerate chemical reactions. Others are transporters, such as hemoglobin, found in red blood cells, which takes oxygen from the lungs to the body's cells.

Proteins are produced from the DNA recipe in two basic steps:

– *Transcription*. Because DNA never leaves the cell's nucleus, a "messenger" must be created to move its information out into the cell. First, a key enzyme called RNA polymerase makes the DNA unwind and "unzip" by breaking the hydrogen bonds between the bases in the paired strands. This process forms two complementary strands—a coding strand and a template strand. RNA (ribonucleic acid), a single-stranded molecule very similar to DNA, is then created as nucleotide bases are strung together by the RNA polymerase in a sequence determined by the DNA template. The new RNA strand has the same information as the original coding strand, with one exception: U (uracil) nucleotides substitute for the Ts in the DNA. The resulting strand is called messenger RNA (mRNA).

– *Translation*. The As, Cs, Gs, and Us in the mRNA are read by the ribosome as three-letter "words," called codons, which are known as the genetic code: AUG GAA UUC UCG CUC. There are 64 codons, representing the 20 or so amino acids that are the building blocks of proteins. These numbers are not equal because sometimes two or more codons code for the same amino acid. There are also "start" and "stop" codons that determine where the protein chain begins and ends.

Now the ribosome has enough information to manufacture, or synthesize, a protein. The ribosome moves along the mRNA strand and reads its sequence, one codon at a time. With the help of another type of RNA called transfer RNA (tRNA), the ribosome adds amino acids one by one to the growing chain, called a polypeptide chain. When it's complete, the chain folds into a specific shape dictated by the amino acid sequence and becomes a protein; its shape determines the protein's function in the body. The translation process from DNA to protein is complete.

Comparing the DNA sequence patterns of humans side-by-side with those of well-studied "model organisms" such as the fruit fly, mouse, pufferfish, and sea squirt is one of the most powerful strategies for identifying human genes and determining how they're regulated and what they do. Conserved sequences—DNA patterns that we share with other organisms—are likely to have important functions or they would have disappeared as the organisms evolved.

Along with helping identify genes and their functions, comparative genomics is shedding light on the functions of the noncoding sequences of DNA found within and between the genes. These segments can regulate gene expression, the process involved in determining when and where in the organism a given gene is turned on or off. Understanding the complex orchestration of gene and protein networks is a crucial aspect of contemporary biomedical research.

We may not look alike, but humans, cows, fish, and microbes have a lot in common. We all retain similar DNA sequences inherited from our shared ancestors who lived hundreds of millions of years ago. Humans share many genes with mice, frogs, flies, and even bacteria and yeast, since their essential functions have been inherited intact over long periods of time.

CENTRAL DOGMA OF MOLECULAR BIOLOGY

While the specific sequence of nucleotides in a DNA molecule can have important information content for a cell, it is actually proteins that do the work of altering a cell's chemistry by acting as biological catalysts called enzymes. In chemistry catalysts are molecules that allow specific chemical reactions to proceed more quickly than they would have otherwise occurred. Catalysts are neither consumed nor altered in the course of such a chemical process and can be used to catalyse the same reaction many times. The term gene is used in many different ways, but one of its narrowest and simplest definitions is that genes spell out the instructions needed to make the enzyme catalysts produced by cells.

The process by which information is extracted from the nucleotide sequence of a gene and then used to make a protein is essentially the same for all living things on Earth and is described by the grandly named central dogma of molecular biology. Information stored in DNA is used to make a more transient, single-stranded polynucleotide called RNA (ribonucleic acid) that is in turn used to make proteins (Figure 1). The process of making an RNA copy of a gene is called transcription and is accomplished through the enzymatic activity of an RNA polymerase. There is a one-to-one correspondence between the nucleotides used to make RNA (G, A, U, and C where "U" is an abbreviation for uracil) and the nucleotide sequences in DNA (G, A, T, and C, respectively).

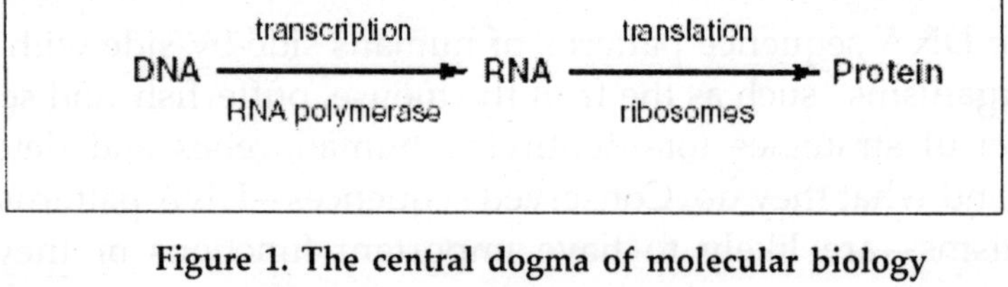

Figure 1. The central dogma of molecular biology

The process of converting that information from nucleotide sequences in RNA to the amino acid sequences that make a protein is called translation and is performed by a complex of proteins and RNA called ribosomes.

GENE STRUCTURE

Formatting and its interpretation are important considerations for any information storage system, be it a written text or a cell's DNA molecule. All cells go about interpreting their genetic instructions in the same way and rely on specific signals to "punctuate" their genes.

Much of the "language" of DNA and the rules of its interpretation were worked out very early in the history of life on earth and, because of their central importance, have changed very little over the course of billions of years. As a result, both prokaryotic (bacteria) and eukaryotic (more complicated organisms like yeast, plants, pets, and people) organisms all use not only the same "alphabet" of nucleotides but also use essentially the same format and approach for storing and utilising their genetic information.

Gene expression, the process of using the information stored in DNA to make an RNA molecule and then a corresponding protein, can have significant energetic and opportunity costs for a cell. Organisms that express unneeded proteins are less likely to survive and reproduce relative to competitors that regulate their gene expression more appropriately. As a result, all cells place particular emphasis on controlling gene expression at its very start by making two crucial distinctions.

First, they must reliably distinguish between those parts of an organism's genome that correspond to the beginnings of genes and those that do not. Second, they must be able to determine which genes code for proteins that are needed at any particular time. Since RNA polymerases are responsible for the initiation of gene expression through their synthesis of RNA copies of genes, it is reasonable that the burden of making those two distinctions falls on them.

Certainly not every nucleotide in a genome can correspond to the start of a gene any more than every letter on a printed page can correspond to the beginning of a sentence with useful information content. By the same token, RNA polymerases cannot simply look for any one particular nucleotide, like A, when looking for the start of a gene because each nucleotide occurs by chance so frequently throughout a cell's DNA. However, particular combinations of nucleotides are not as likely to occur by chance, and the greater the number of nucleotides involved, the smaller a chance occurrence becomes.

The probability (P) that a string of nucleotides will occur by chance alone can be determined by the relatively simple formula $P = (1/4)n$ if all nucleotides are present at the same frequency and where n is the string's length. Prokaryotic RNA polymerases actually scan along DNA looking for a specific set of approximately 13 nucleotides (1 nucleotide that serves as a transcriptional start site, 6 that are 10 nucleotides 5' to the start site, and 6 more that are 35 nucleotides 5' to it) that mark the beginning of genes. Those nucleotides, taken as a whole and in the proper positions relative to each other, are called promoter sequences.

Given that most prokaryotic genomes are only a few million nucleotides long, these promoter sequences, which should occur only by chance about once in every 70 million nucleotides, allow RNA polymerases to uniquely identify the beginnings of genes with great statistical confidence. Eukaryotic genomes tend to be several orders of magnitude larger than those of prokaryotes and, as a result, eukaryotic RNA polymerases tend to recognise larger and more complex promoter sequences so that they too can reliably recognise the beginning of genes.

Two French biochemists, F. Jacob and J. Monod, were the first to obtain direct molecular insights into how cells distinguish between genes that should be transcribed and those that should not. Their work on prokaryotic gene regulation revealed that the expression of structural genes (those that code for proteins involved in cell structure or metabolism) was controlled by specific regulatory genes. The proteins encoded by these regulatory genes are typically capable of binding to a cell's DNA near the promoter of the genes whose expression they control in some circumstances but not in others. It is the ability of these regulatory proteins to bind or not bind to specific nucleotide sequences in a fashion that is dependent on their ability to sense a cell's chemical environment that allows living things to respond appropriately to their environment.

When the binding of these proteins makes it easier for an RNA polymerase to initiate transcription, positive regulation is said to have occurred. Negative regulation describes those situations where binding of the regulatory protein prevents transcription from occurring. Eventually, most prokaryotic structural genes were found to be turned on or off by just one or two regulatory proteins. Eukaryotes, with their substantially more complicated genomes and transcriptional needs, use larger numbers (usually seven or more) and combinations of regulatory proteins to control the expression of their structural genes.

Genetic Code

While nucleotides are the building blocks that cells use to make their information storage and transfer molecules (DNA and RNA, respectively), amino acids are the units that are strung together to make the proteins that actually do most of the work of altering a cell's chemical environment. The function of a protein is intimately dependent on the order in which its amino acids are linked by ribosomes during translation and, as has already been discussed, that order is determined by the instructions transcribed into RNA molecules by RNA polymerases. However, although only four different nucleotides (nt) are used to make DNA and RNA molecules, 20 different amino acids (each with its own distinctive chemistry) are used in protein synthesis. There cannot be a simple one-to-one correspondence between the nucleotides of genes and the amino acids of the proteins they encode. The 16 different possible pairs of nucleotides also fall short of the task (2 nt $\neq$ 1 aa; $4^2 < 20$). However, the four nucleotides can be arranged in a total of 64 different combinations of three ($4^3 = 64$). As a result, it is necessary for ribosomes to use a triplet code to translate the information in DNA and RNA into the amino acid sequence of proteins. With only three exceptions, each group of three nucleotides (a codon) in an RNA copy of the coding portion of a gene corresponds to a specific amino acid.

The three codons that do not instruct ribosomes to insert a specific amino acid are called stop codons (functionally equivalent to a period at the end of a sentence) because they cause translation to be terminated. This same genetic code seems to have been in place since the earliest history of life on earth and, with only a few exceptions, is universally used by all living things today.

Each amino acid can be assigned to one of essentially four different categories: nonpolar, polar, positively charged, and negatively charged. A single change within a triplet codon is usually not sufficient to cause a codon to code for an amino acid in a different group. In short, the genetic code is remarkably robust and minimises the extent to which mistakes in the nucleotide sequences of genes can change the functions of the proteins they encode.

Translation by ribosomes starts at translation-initiation sites on RNA copies of genes and proceeds until a stop codon is encountered. Just as three codons of the genetic code are reserved as stop codons, one triplet codon is always used as a start codon. Specifically, the codon AUG is used both to code for the amino acid methionine as well as to mark the precise spot along an RNA molecule where translation begins in both prokaryotes and eukaryotes. Accurate translation can only occur when ribosomes examine codons in the phase or reading frame that is established by a gene's start codon.

Unless a mistake involving some multiple of three nucleotides occurs, alterations of a gene's reading frame change every amino acid coded downstream of the alteration, and such alterations typically result in the production of a truncated version of the protein due to ribosomes encountering a premature stop codon.

Most genes code for proteins that are hundreds of amino acids long. Since stop codons occur in a randomly generated sequence at about every 20th triplet codon (3 codons out of 64), one of the reading frames of the RNA copies of most genes has unusually long runs of codons in which no stop codons occur. These strings of codons uninterrupted by stop codons are known as open reading frames (ORFs) and are a distinguishing feature of many prokaryotic and eukaryotic genes.

Introns and Exons

The messenger RNA (mRNA) copies of prokaryotic genes correspond perfectly to the DNA sequences present in the organism's genome with the exception that the nucleotide uracil (U) is used in place of thymine (T). In fact, translation by ribosomes almost always begins while RNA polymerases are still actively transcribing a prokaryotic gene. Eukaryotic RNA polymerases also use uracil in place of thymine, but much more striking differences are commonly found between the mRNA molecules seen by ribosomes and the nucleotide sequences of the eukaryotic genes that code for them.

In eukaryotes the two steps of gene expression are physically separated by the nuclear membrane, with transcription occurring exclusively within the nucleus and translation occurring only after mRNAs have been exported to the cytoplasm. As a result, the RNA molecules transcribed by eukaryotic RNA polymerases can be modified before ribosomes ever encounter them. The most dramatic modification that is made to the primary RNA transcripts of most eukaryotic genes is called splicing and involves the precise excision of internal sequences known as introns and the rejoining of the exons that flank them (Figure 2). Splicing is far from a trivial process and most eukaryotic genes have a large number of sometimes very large introns.

An extreme example is the gene associated with the disease cystic fibrosis in humans, which has 24 introns and is over 1 million nucleotides (1 mega base pair or 1 Mb) long even though the mRNA seen by ribosomes is only about 1,000 nucleotides (1 kilo base pair or 1 kb) long. Failure to appropriately splice the introns out of a primary eukaryotic RNA transcript typically introduces frame shifts or premature stop codons that render useless any protein translated by a ribosome.

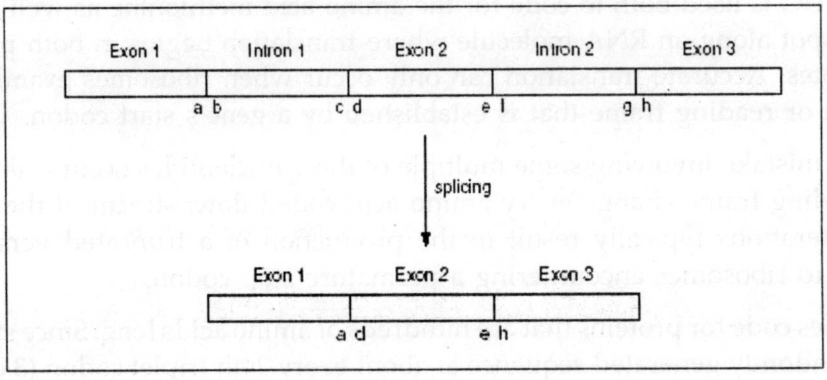

Figure 2. Splicing of primary transcripts of eukaryotic genes results in removal of introns and the precise joining of exons

Regardless of the tissue or even the organism being considered, the vast majority of eukaryotic introns conform to what is known as the "GT-AG rule," meaning that the first two nucleotides in the DNA sequence of all introns begin with the dinucleotide GT and end with the dinucleotide AG.

Pairs of nucleotides occur too often just by chance to be a sufficient signal for the enzyme complexes responsible for splicing in eukaryotes, spliceosomes, and approximately six additional nucleotides at the 5' and 3' ends of introns are also scrutinised-sometimes differently in some cell types relative to others. This alternative splicing allows a huge increase in the diversity of proteins that eukaryotic organisms can use and is accomplished by often subtle modifications of spliceosomes and accessory proteins that are responsible for recognising intron/exon boundaries.

GENOME ORGANISATION

The genome of a typical bacterium comes as a single DNA molecule, which, if extended, would be about 2 mm long. The DNA of higher organisms is organised into chromosomes-normal human cells contain 23 chromosome pairs. The total amount of genetic information per cell-the sequence of nucleotides of DNA-is very nearly constant for all members of a species, but varies widely between species.

Organism	Genome size (base pairs)
Epstein-Barr virus	0.172×10^6
Bacterium (E. coli)	4.6×0^6
Yeast (S. cerevisiae)	12.5×10^6
Nematode worm (C. elegans)	100.3×10^6
Thale cress (A. thaliana)	115.4×10^6
Fruit fly (D. melanogaster)	128.3×10^6
Human (H. sapiens)	3223×10^6

Not all DNA codes for proteins. Conversely, some genes exist in multiple copies. Therefore the amount of protein sequence information in a cell cannot easily be estimated from the genome size.

A single gene coding for a particular protein corresponds to a sequence of nucleotides along one or more regions of a molecule of DNA. The DNA sequence is collinear with the protein sequence. In species for which the genetic material is double-stranded DNA, genes may appear on either strand. Bacterial genes are continuous regions of DNA.

Therefore the functional unit of genetic sequence information from a bacterium is a string of 3N nucleotides encoding a string of N amino acids, or a string of N nucleotides encoding a structural RNA molecule. Such a string, equipped with annotations, would form a typical entry in one of the genetic sequence archives. In eukaryotes the nucleotide sequences that encode the amino acid sequences of individual proteins are organised in a more complex manner.

Frequently one gene appears split into separated segments in the genomic DNA. An exon (expressed region) is a stretch of DNA retained in the mature messenger RNA that a ribosome translates into protein. An intron is an intervening region between two exons. Cellular machinery splices together the proper segments, in RNA transcripts, based on signal sequences flanking the exons in the sequences themselves. Many introns are very long-in some cases substantially longer than the exons.

Control information organises the expression of genes. Control mechanisms may turn genes on and off (or more finely regulate gene expression) in response to concentrations of nutrients, or to stress, or to unfold complex programs of development during the lifetime of the organism. Many control regions of DNA lie near the segments coding for proteins. They contain signal sequences that serve as binding sites for the molecules that transcribe the DNA sequence, or sequences that bind regulatory molecules that can block transcription.

Bacterial genomes contain examples of contiguous genes coding for several proteins that catalyse successive steps in an integrated sequence of reactions, all under the control of the same regulatory sequence. One can readily understand the utility of a parallel control mechanism. In animals, methylation of DNA provides the signals for tissuespecific expression of developmentally regulated genes.

Products of certain genes cause cells to commit suicide-a process called apoptosis. Defects in the apoptotic mechanism leading to uncontrolled growth are observed in some cancers, and stimulation of these mechanisms is a general approach to cancer therapy. Reverse engineering the sequence is complicated by the fact that the resulting image of the mass-storage device will not be a file-by-file copy, but rather a streaming dump of the bytes in the order they were entered into the device.

Furthermore, the files are known to be fragmented. In addition, some of the device contains erased files or other garbage. Once the garbage has been recognised and discarded and the fragmented files reassembled, the reverse engineering of the codes can be undertaken with only a partial, and sometimes incorrect, understanding of the CPU on which the codes run. In fact, deducing the structure and function of the CPU is part of the project, since some of the 3.2 gigabytes are the binary specifications for the computer-assisted-manufacturing process that fabricates the CPU. In addition, one must also consider that the huge database also contains code generated from the result of literally millions of maintenance revisions performed by the worst possible set of kludge-using, spaghetti-coding, opportunistic hackers who delight in clever tricks like writing self-modifying code and relying upon undocumented system quirks.

PROTEOME PROJECT

The proteome project is a large-scale programme dealing in an integral way with patterns of expression of proteins in biological systems, in ways that complement and extend genome projects. The basic goal is a spatio-temporal description of the deployment of proteins in the organism. The rates of synthesis of different proteins vary among different tissues and different cell types and states of activity.

Methods are available for analysis of transcription patterns of genes. However, because proteins 'turn over' at different rates, it is also necessary to measure proteins directly. High-resolution two-dimensional polyacylamide gel electrophoresis (2D PAGE) shows the pattern of protein content in a sample. Mass-spectroscopic techniques identify the proteins into which the sample has been separated. In principle, a database of amino acid sequences of proteins is inherent in the database of nucleotide sequences of DNA, by virtue of the genetic code. Indeed, new protein sequence data are now being determined almost exclusively by translation of DNA sequences, rather than by direct sequencing of proteins.

First, we must assume that it is possible correctly to identify within the DNA data stream the regions that encode proteins. The pattern-recognition programs that address this question are subject to three types of errors: a genuine protein sequence may be missed entirely, or an incomplete protein may be reported, or a gene may be incorrectly spliced.

Several variations on the theme add to the complexity: Genes for different proteins may overlap, or genes may be assembled from exons in different ways in different tissues. Conversely, some genetic sequences that appear to code for proteins may in fact be defective or not expressed. A protein inferred from a genome sequence is a hypothetical object until an experiment verifies its existence.

Second, in many cases the expression of a gene produces a molecule that must be modified within a cell, to make a mature protein that differs significantly from the one suggested by translation of the gene sequence. In many cases the missing details of post-translational modifications-the molecular analogues of body piercing-are quite important. Post-translational modifications include addition of ligands (for instance the covalently-bound haem group of cytochrome c), glycosylation, methylation, excision of peptides, and many others.

Patterns of disulphide bridges-primary chemical bonds between cysteine residues-cannot be deduced from the amino acid sequence. In some cases, mRNA is edited before translation, creating changes in amino acid sequences that are not inferrable from the genes.

TRANSMISSION OF GENETIC INFORMATION

How hereditary information is stored, passed on, and implemented is perhaps the fundamental problem of biology. Three types of maps have been essential.

- Linkage maps of genes
- Banding patterns of chromosomes
- DNA sequences

Linkage Maps of Genes

A gene map is classically determined by observed patterns of heredity. Linkage groups and recombination frequencies can detect whether genes are on the same or different chromosomes, and, for genes on the same chromosome, how far apart they are. The principle is that the farther apart two linked genes are, the more likely they are to recombine, by crossing over during meiosis. Indeed, two genes on the same chromosome but very far apart will appear to be unlinked. The unit of length in a gene map is the Morgan, defined by the relation that 1 cM corresponds to a 1% recombination frequency.

Chromosome Banding Pattern Maps

Chromosomes are physical objects. Banding patterns are visible features on them. The nomenclature is as follows: In many organisms, chromosomes are numbered in order of size, 1 being the largest. The two arms of human chromosomes, separated by the centromere, are called the p (petite = short) arm and q (= queue) arm. Regions within the chromosome are numbered p1, p2, . . . and q1, q2 . . . outward from the centromere. Subsequent digits indicate subdivisions of bands.

For example, certain bands on the q arm of human chromosome 15 are labelled 15q11.1, 15q11.2, 15q12. Originally bands 15q11 and 15q12 were defined; subsequently 15q11 was divided into 15q11.1 and 15q11.2. Deletions including this region are associated with Prader-Willi and Angleman syndromes. These syndromes have the interesting feature that the alternative clinical consequences depend on whether the affected chromosome is paternal or maternal. This observation of genomic imprinting shows that the genetic information in a fertilised egg is not simply the bare DNA sequences contributed by the parents.

Chromosomes of paternal and maternal origin have different states of methylation, signals for differential expression of their genes. The process of modifying the DNA which takes place during differentiation in development is already present in the zygote.

DNA Sequence

Physically a sequence of nucleotides in the molecule, computationally a string of characters A, T, G and C. Genes are regions of the sequence, in many cases interrupted by non-coding regions. These represent three very different types of data. Genes, as discovered by Mendel, were entirely abstract entities. Chromosomes are physical objects, the banding patterns their visible landmarks. Only with DNA sequences are we dealing directly with stored hereditary information in its physical form.

It was the very great achievement of the last century of biology to forge connections between these three types of data. The first steps-and giant strides they were indeed-proved that, for any chromosome, the maps are one-dimensional arrays, and indeed that they are collinear. Any schoolchild now knows that genes are strung out along chromosomes, and that each gene corresponds to a DNA sequence. But the proofs of these statements earned a large number of Nobel prizes.Splitting a long molecule of DNA-for example, the DNA in an entire chromosome -into fragments of convenient size for cloning and sequencing requires additional maps to report the order of the fragments, so that the entire sequence can be reconstructed from the sequences of the fragments. A restriction endonuclease is an enzyme that cuts DNA at a specific sequence, usually about 6 bp long. Cutting DNA with several restriction enzymes with different specificities produces sets of overlapping fragments.

From the sizes of the fragments it is possible to construct a restriction map, stating the order and distance between the restriction enzyme cleavage sites. A mutation in one of these cleavage sites will change the sizes of the fragments produced by the corresponding enzyme, allowing the mutation to be located in the map. Restriction enzymes can produce fairly large pieces of DNA. Cutting the DNA into smaller pieces, which are cloned and ordered by sequence overlaps produces a finer dissection of the DNA called a contig map.

In the past, the connections between chromosomes, genes and DNA sequences have been essential for identifying the molecular deficits underlying inherited diseases, such as Huntington disease or cystic fibrosis. Sequencing of the human genome has changed the situation radically.

Given a disease attributable to a defective protein:

- If we know the protein involved, we can pursue rational approaches to therapy.
- If we know the gene involved, we can devise tests to identify sufferers or carriers.
- In many cases, knowledge of the chromosomal location of the gene is unnecessary for either therapy or detection; it is required only for identifying the gene, providing a bridge between the patterns of inheritance and the DNA sequence.

For instance, in the case of sickle-cell anaemia, we know the protein involved. The disease arises from a single point mutation in haemoglobin. We can proceed directly to drug design. We need the DNA sequence only for genetic testing and counselling. In contrast, if we know neither the protein nor the gene, we must somehow go from the phenotype back to the gene, a process called positional cloning or reverse genetics. Positional cloning used to involve a kind of 'Tinker to Evers to Chance' cascade from the gene map to the chromosome map to the DNA sequence.

Patterns of inheritance identify the type of genetic defect responsible for a condition. They show, for example, that Huntington disease and cystic fibrosis are caused by single genes. To find the gene associated with cystic fibrosis it was necessary to begin with the gene map, using linkage patterns of heredity in affected families to localise the affected gene to a particular region of a particular chromosome.

Knowing the general region of the chromosome, it was then possible to search the DNA of that region to identify candidate genes, and finally to pinpoint the particular gene responsible and sequence it. In contrast, many diseases do not show simple inheritance, or, even if only a single gene is involved, heredity creates only a predisposition, the clinical consequences of which depend on environmental factors. The full human genome sequence, and measurements of expression patterns, will be essential to identify the genetic components of these more complex cases.

Mappings Genomes

A gene linkage map can be calibrated to chromosome banding patterns through observation of individuals with deletions or translocations of parts of chromosomes. The genes responsible for phenotypic changes associated with a deletion must lie within the deletion. Translocations are correlated with altered patterns of linkage and recombination.

There have been several approaches to coordinating chromosome banding patterns with individual DNA sequences of genes:

— In Fluorescent In Situ Hybridization (FISH) a probe sequence is labelled with fluorescent dye. The probe is hybridized with the chromosomes, and the chromosomal location where the probe is bound shows up directly in a photograph.

  Typical resolution is ~105 base pairs, but specialised new techniques can achieve high resolution, down to 1 kbp. Simultaneous FISH with two probes can detect linkage and even estimate genetic distances. This is important in species for which the generation time is long enough to make standard genetic approaches inconvenient. FISH can also detect chromosomal abnormalities.

— Somatic cell hybrids are rodent cells containing few, one, or even partial human chromosomes. Hybridization of a probe sequence with a panel of somatic cell hybrids, detected by fluorescence, can identify which chromosome contains the probe. This approach has been superseded by use of clones of yeast or bacteria or phage containing fragments of human DNA in artificial chromosomes (YACs and BACs and PACs).

Of course, for fully-sequenced genomes these methods are obsolete. Given a DNA sequence one would just look it up (but mind the gaps).

High-resolution Maps

Formerly, genes were the only visible portions of genomes. Now, markers are no longer limited to genes with phenotypically observable effects, which are anyway too sparse for an adequately high-resolution map of the human genome. Now that we can interrogate DNA sequences directly, any features of DNA that vary among individuals can serve as markers, including:

- Variable number tandem repeats (VNTRs), also called minisatellites. VNTRs contain regions 10-100 base pairs long, repeated a variable number of times-same sequence, different number of repeats. In any individual, VNTRs based on the same repeat motif may appear only once in the genome; or several times, with different lengths on different chromosomes.

  The distribution of the sizes of the repeats is the marker. Inheritance of VNTRs can be followed in a family and mapped to a disease phenotype like any other trait. VNTRs were the first genetic sequence data used for personal identification-genetic fingerprints-in paternity and in criminal cases.

  Formerly, VNTRs were observed by producing restriction fragment length polymorphisms (RFLPs) from them. VNTRs are generally flanked by recognition sites for the same restriction enzyme, which will neatly excise them.

  The results can be spread out on a gel, and detected by Southern blotting. Note the distinction: VNTRs are characteristics of genome sequences; RFLPs are artificial mixtures of short stretches of DNA created in the laboratory in order to identify VNTRs. It is much easier and more efficient to measure the sizes of VNTRs by amplifying them with PCR, and this method has replaced the use of restriction enzymes.

- Short tandem repeat polymorphisms (STRPs), also called microsatellites. STRPs are regions of only 2-5 bp but repeated many times; typically 10-30 consecutive copies. They have several advantages as markers over VNTRs, one of which is a more even distribution over the human genome. There is no reason why these markers need lie within expressed genes, and usually they do not.

  Panels of microsatellite markers greatly simplify the identification of genes. It is interesting to compare a recent project to identify a disease gene now that the human genome sequence is available, with such classic studies as the identification of the gene for cystic fibrosis. Additional mapping techniques deal more directly with the DNA sequences, and can short-circuit the process of gene identification:

- A contig, or contiguous clone map, is a series of overlapping DNA clones of known order along a chromosome from an organism of interest-for instance, human-stored in yeast or bacterial cells as YACs (Yeast Artificial Chromosomes) or BACs (Bacterial Artificial Chromosomes). A contig map can produce a very fine mapping of a genome. In a YAC, human DNA is stably integrated into a small extra chromosome in a yeast cell. A YAC can contain up to 106 base pairs.

  In principle, the entire human genome could be represented in 10 000 YAC clones. In a BAC, human DNA is inserted into a plasmid in an E. coli cell. A BAC can carry about 250 000 bp. Despite their smaller capacities, BACs are preferred to YACs because of their greater stability and ease of handling.

— A sequence tagged site (STS) is a short, sequenced region of DNA, typically 200-600 bp long, that appears in a unique location in the genome. It need not be polymorphic. An STS can be mapped into the genome by using PCR to test for the presence of the sequence in the cells containing a contig map.

  One type of STS arises from an expressed sequence tag (EST), a piece of cDNA (complementary DNA; that is, a DNA sequence derived from the messenger RNA of an expressed gene). The sequence contains only the exons of the gene, spliced together to form the sequence that encodes the protein. cDNA sequences can be mapped to chromosomes using FISH, or located within contig maps.

If you are working with an organism for which the full-genome sequence is not known, but for which full contig maps are available for all chromosomes, you would identify STS markers tightly linked to your gene, then locate these markers in the contig maps.

GENES IN GENOMES

Computer programs for genome analysis identify open reading frames or ORFs. An ORF is a region of DNA sequence that begins with an initiation codon (ATG) and ends with a stop codon. An ORF is a potential protein-coding region. Approaches to identifying protein-coding regions choose from or combine two possible approaches:

— *Detection of regions similar to known coding regions from other organisms*: These regions may encode amino acid sequences similar to known proteins, or may be similar to Expressed Sequence Tags (ESTs). Because ESTs are derived from messenger RNA, they correspond to genes known to be transcribed. It is necessary to sequence only a few hundred initial bases of cDNA to give enough information to identify a gene: characterisation of genes by ESTs is like indexing poems or songs by their first lines.

— Ab initio methods, that seek to identify genes from the properties of the DNA sequences themselves. Computer-assisted annotation of genomes is more complete and accurate for bacteria than for eukaryotes. Bacterial genes are relatively easy to identify because they are contiguous-they lack the introns characteristic of eukaryotic genomes-and the intergene spaces are small. In higher organisms, identifying genes is harder. Identification of exons is one problem, assembling them is another. Alternative splicing patterns present a particular difficulty.

A framework for ab initio gene identification in eukaryotic genomes includes the following features:

— The initial (5′) exon starts with a transcription start point, preceded by a core promotor site such as the TATA box typically ~30 bp upstream. It is free of in-frame stop codons, and ends immediately before a GT splice signal.

— Internal exons, like initial exons, are free of in-frame stop codons. They begin immediately after an AG splice signal and end immediately before a GT splice signal.

— The final (3′) exon starts immediately after an AG splice signal and ends with a stop codon, followed by a polyadenylation signal sequence.

All coding regions have nonrandom sequence characteristics, based partly on codon usage preferences. Empirically, it is found that statistics of hexanucleotides perform best in distinguishing coding from noncoding regions. Starting from a set of known genes from an organism as a training set, pattern recognition programs can be tuned to particular genomes. Accurate gene detection is a crucial component of genome sequence analysis.

GENOMES OF PROKARYOTES

Most prokaryotic cells contain their genetic material in the form of a large single circular piece of double-stranded DNA, usually less than 5 Mbp long. In addition they may contain plasmids. The protein-coding regions of bacterial genomes do not contain introns. In many prokaryotic genomes the protein-coding regions are partially organised into operons-tandem genes transcribed into a single messenger RNA molecule, under common transcriptional control. In bacteria, the genes of many operons code for proteins with related functions.

For instance, successive genes in the trp operon of E. coli code for proteins that catalyse successive steps in the biosynthesis of tryptophan. In archaea, a metabolic relationship between genes in operons is less frequently observed.

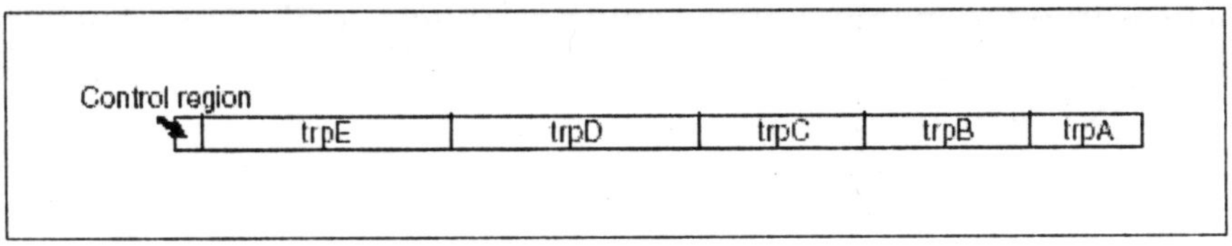

Figure 3. The trp operon in E. coli begins with a control region containing promoter, operator and leader sequences

Five structural genes encode proteins that catalyse successive steps in the synthesis of the amino acid tryptophan from its precursor chorismate:

— *Reaction step (1)*: trpE and trpD encode two components of anthranilate synthase. This tetrameric enzyme, comprising two copies of each subunit, catalyses the conversion of chorismate to anthranilate.

— *Reaction step (2)*: The protein encoded by trpD also catalyses the subsequent phosphoribosylation of anthranilate.

— *Reaction step (3)*: trpC encodes another bifunctional enzyme, phosphoribosyl-anthranilate isomerase - indoleglycerolphosphate synthase. It converts phosphoribosyl anthranilate to indoleglycerol- phosphate, through the intermediate, carboxy-phenylaminodeoxyribulose phosphate.

— *Reaction steps (4) and (5)*: trpB and trpA encode the β and α subunits, respectively, of a third bifunctional enzyme, tryptophan synthase (an $\alpha_2\beta_2$ tetramer). A tunnel within the structure of this enzyme delivers, without release to the solvent, the intermediate produced by the α subunit - indoleglycerolphosphate → indole - to the active site of the β subunit - which converts indole → tryptophan.

A separate gene, trpR, not closely linked to this operon, codes for the trp repressor. The repressor can bind to the operator sequence in the DNA (within the control region) only when binding tryptophan. Binding of the repressor blocks access of RNA polymerase to the promoter, turning the pathway off when tryptophan is abundant.

Further control of transcription in response to tryptophan levels is exerted by the attenuator element in the mRNA, within the leader sequence. The attenuator region

— contains two tandem trp codons and

— can adopt alternative secondary structures, one of which terminates transcription.

Levels of tryptophan govern levels of trp-tRNAs, which govern the rate of progress of the tandem trp codons through the ribosome. Stalling on the ribosome at the tandem trp codons in response to low tryptophan levels reduces the formation of the mRNA secondary structure that terminates transcription.

The typical prokaryotic genome contains only a relatively small amount of noncoding DNA (in comparison with eukaryotes), distributed throughout the sequence. In E. coli only ~11% of the DNA is noncoding.

Genome of Bacterium Escherichia Coli

E. coli, strain K-12, has long been the workhorse of molecular biology. The genome of strain MG1655, published in 1997 by the group of F. Blattner at the University of Wisconsin, contains 4 639 221 bp in a single circular DNA molecule, with no plastids. Approximately 89% of the sequence codes for proteins or Control region trpE trpD trpC trpB trpA structural RNAs. An inventory reveals:

— 4284 protein-coding genes

— 122 structural RNA genes

— noncoding repeat sequences

— regulatory elements

— transcription/translation guides

— transposases

- prophage remnants
- insertion sequence elements
- patches of unusual composition, likely to be foreign elements introduced by horizontal transfer.

Analysis of the genome sequence required identification and annotation of protein-coding genes and other functional regions. Many E. coli proteins were known before the sequencing was complete, from the many years of intensive investigation: 1853 proteins had been described before publication of the genome sequence. Other genes could be assigned functions from identification of homologues by searching in sequence databanks.

The narrower the range of specificity of the function of the homologues, the more precise could be the assignment. Currently, over 80% of proteins can be assigned at least a general function. Other regions of the genome are recognised as regulatory sites, or mobile genetic elements, also on the basis of similarity to homologous sequences known in other organisms.

The distribution of protein-coding genes over the genome of E. coli does not seem to follow any simple rules, either along the DNA or on different strands. Indeed, comparison of strains suggests that genes are mobile. The E. coli genome is relatively gene dense. Genes coding for proteins or structural RNAs occupy ~89% of the sequence. The average size of an open reading frame is 317 amino acids. If the genes were evenly distributed, the average intergenic region would be 130 bp; the observed average distance between genes is 118 bp.

However, the sizes of intergenic regions vary considerably. Some intergenic regions are large. These contain sites of regulatory function, and repeated sequences. The longest intergenic region, 1730 bp, contains noncoding repeat sequences. Approximately three-quarters of the transcribed units contain only one gene; the rest contain several consecutive genes, or operons. It is estimated that the E. coli genome contains 630-700 operons. Operons vary in size, although few contain more than five genes. Genes within operons tend to have related functions. In some cases, the same DNA sequence encodes parts of more than one polypeptide chain. One gene codes for both the τ and γ subunits of DNA polymerase III. Translation of the entire gene forms the subunit.

The γ subunit corresponds approximately to the N-terminal two-thirds of the subunit. A frameshift on the ribosome at this point leads to chain termination 50% of the time, causing a 1:1 ratio of expressed τ and γ subunits. There do not appear to be any overlapping genes in which different reading frames both code for expressed proteins.

In other cases, the same polypeptide chain appears in more than one enzyme. A protein that functions on its own as lipoate dehydrogenase is also an essential subunit of pyruvate dehydrogenase, 2-oxoglutarate dehydrogenase and the glycine cleavage complex.Having the complete genome, we can examine the protein repertoire of E. coli. The largest class of proteins are the enzymes-appoximately 30% of the total genes. Many enzymatic functions are shared by more than one protein. Some of these sets of functionally similar enzymes are very closely related, and appear to have arisen by duplication, either in E. coli itself or in an ancestor. Other sets of functionally similar enzymes have very dissimilar sequences, and differ in specificity, regulation, or intracellular location.

Several features of E. coli's generous endowment of enzymes give it an extensive and versatile metabolic competence, which allow it to grow and compete under varying conditions:

- It can synthesise all components of proteins and nucleic acids (amino acids and nucleotides), and cofactors.
- It has metabolic flexibility: Both aerobic and anaerobic growth are possible, utilising different pathways of energy capture. It can grow on many different carbon sources. Not all metabolic pathways are active at any time, the versatility allows response to changes in conditions.
- Even for specific metabolic reactions there are many cases of multiple enzymes. These provide redundancy, and contribute to an ability to tune metabolism to varying conditions.
- However, E. coli does not possess a complete range of enzymatic capacity. It cannot fix CO_2 or N_2.

Here described some of the static features of the E. coli genome and its protein repertoire. Current research is moving into dynamic aspects, to investigations of protein expression patterns in time and space.

Archaeon Methanococcus Jannaschii

S. Luria once suggested that to determine common features of all life one should not try to survey everything, but rather identify the organism most different from us and see what we have in common with it. The assumption was that the way to do this would be to find an organism adapted to the most different environment.

Deep-sea exploration has revealed environments as far from the familiar as those portrayed in science fiction. Hydrothermal vents are underwater volcanoes emitting hot lava and gases through cracks in the ocean floor. They create niches for communities of living things disconnected from the surface, depending on the minerals exuded from the vent for their sources of inorganic nutrients. They support living communities of micro-organisms that are the only known forms of life not dependent on sunlight, directly or indirectly, as their energy source.

The micro-organism Methanococcus jannaschii was collected from a hydrothermal vent 2600 m deep off the coast of Baja California, Mexico, in 1983. It is a thermophilic organism, surviving at temperatures from 48-94°C, with an optimum at 85°C. It is a strict anaerobe, capable of self-reproduction from inorganic components. Its overall metabolic equation is to synthesise methane from H_2 and CO_2.

M. jannaschii belongs to the archaea, one of the three major divisions of life along with the bacteria and eukarya. The archaea comprise groups of prokaryotes, including organisms adapted to extreme environmental conditions such as high temperature and pressure, or high salt concentration. The genome of M. jannaschii was sequenced in 1996 by The Institute for Genomic Research (TIGR). It was the first archaeal genome sequenced. It contains a large chromosome containing a circular double-stranded DNA molecule 1664 976 bp long, and two extrachromosomal elements of 58 407 and 16 550 bp respectively. There are 1743 predicted coding regions, of which 1682 are on the chromosome, and 44 and 12 are on the large and small extrachromosomal elements respectively. Some RNA genes contain introns. As in other prokaryotic genomes there is little noncoding DNA.

M. jannaschii would appear to satisfy Luria's goal of finding our most distant extant relative. Comparison of its genome sequence with others shows that it is distantly related to other forms of life. Only 38% of the open reading frames could be assigned a function, on the basis of homology to proteins known from other organisms. However, to everyone's great surprise, archaea are in some ways more closely related to eukaryotes than to bacteria! They are a complex mixture. In archaea, proteins involved in transcription, translation and regulation are more similar to those of eukaryotes. Archaeal proteins involved in metabolism are more similar to those of bacteria.

Genome of one of the simplest organisms: Mycoplasma genitalium

Mycoplasma genitalium is an infectious bacterium, the cause of nongonococcal uretheitis. Its genome was sequenced in 1995 by a collaboration of groups at TIGR, The Johns Hopkins University and The University of North Carolina. The genome is a single DNA molecule containing 580 070 base pairs. This is the smallest cellular genome yet sequenced. So far, M. genitalium is the closest we have to a minimal organism, the smallest capable of independent life.The genome is dense in coding regions. 468 genes have been identified as expressed proteins. Some regions of the sequence are gene rich, others gene poor, but overall 85% of the sequence is coding. The average length of a coding region is 1040 bp. As in other bacterial genomes, the coding regions do not contain introns. Further compression of the genome is achieved by overlapping genes. It appears that many of these have arisen through loss of stop codons.

The gene repertoire of M. genitalium includes some that encode proteins essential for independent reproduction, such as those involved in DNA replication, transcription, and translation, plus ribosomal and transfer RNAs. Other genes are specific for the infectious activity, including adhesins that mediate binding to infected cells, other molecules for defence against the host's immune system, and many transport proteins. As adaptation to the parasitic lifestyle of the organism, there has been widespread loss of metabolic enzymes, including those responsible for amino acid biosynthesis-indeed, one amino acid is absent from all M. genitalium proteins.

GENOMES OF EUKARYOTES

It is rare in science to encounter a completely new world containing phenomena entirely unsuspected. The complexity of the eukaryotic genome is such a world. In eukaryotic cells, the majority of DNA is in the nucleus, separated into bundles of nucleoprotein, the chromosomes. Each chromosome contains a single double-stranded DNA molecule.

Smaller amounts of DNA appear in organelles-mitochondria and chloroplasts. The organelles originated as intracellular parasites. Organelle genomes usually have the form of circular doublestranded DNA, but are sometimes linear and sometimes appear as multiple circles. The genetic code by which organelle genes are translated differs from that of nuclear genes.

Nuclear genomes of different species vary widely in size. The correlation between genome size and complexity of the organism is very rough. It certainly does not support any preconception that humans stand on a pinnacle. In many cases differences in genome size reflect different amounts of simple repetitive sequences.

In addition to variation in DNA content, eukaryotic species vary in the number of chromosomes and distribution of genes among them. Some differences in the distribution of genes among chromosomes involve translocations, or chromosome fragmentations or joinings. For instance, humans have 23 pairs of chromosomes; chimpanzees have 24. Human chromosome 2 is equivalent to a fusion of chimpanzee chromosomes 12 and 13.

The difficulty of chromosome pairing during mitosis in a zygote after such an event can contribute to the reproductive isolation required for species separation. Other differences in chromosome complement reflect duplication or hybridization events. The wheat first used in agriculture, in the Middle East at least 10 000-15 000 years ago, is a diploid called einkorn (Triticum monococcum), containing 14 pairs of chromosomes. Emmer wheat (T. dicoccum), also cultivated since palaeolithic times, and durum wheat (T. turgidum), are merged hybrids of relatives of einkorn with other wild grasses, to form tetraploid species.

Additional hybridizations, to different wild wheats, gave hexaploid forms, including spelt (T. spelta), and modern common wheat T. aestivum. Triticale, a robust crop developed in modern agriculture and currently used primarily for animal feed, is an artificial genus arising from crossing durum wheat (Triticum turgidum) and rye (Secale cereale). Most triticale varieties are hexaploids.

Variety of wheat	Classification	Chromosome complement
einkorn	Triticum monococcum	AA
emmer wheat	Triticum dicoccum	AABB
durum wheat	Triticum turgidum	AABB
spelt	Triticum spelta	AABBDD
common wheat	Triticum aestivum	AABBDD
triticale	Triticosecale	AABBRR

A = genome of original diploid wheat or a relative,

B = genome of a wild grass Aegilops speltoides or Triticum speltoides or a relative,

D = genome of another wild grass, Triticum tauschii or a relative,

R = genome of rye Secale cereale.

All these species are still cultivated-some to only minor extents-and have their individual uses in cooking. Spelt, or farro in Italian, is the basis of a well-known soup; pasta is made from durum wheat; and bread from T. aestivum. Even within single chromosomes, gene families are common in eukaryotes.

Some family members are paralogues-related genes that have diverged to provide separate functions in the same species. Other related sequences may be pseudogenes, which may have arisen by duplication, or by retrotransposition from messenger RNA, followed by the accumulation of mutations to the point of loss of function. The human globin gene cluster is a good example.

Genome of Saccharomyces Cerevisiae

Yeast is one of the simplest known eukaryotic organisms. Its cells, like our own, contain a nucleus and other specialised intracellular compartments. The sequencing of its genome, by an unusually effective international consortium involving ~100 laboratories, was completed in 1992. The yeast genome contains 12 057 500 bp of nuclear DNA, distributed over 16 chromosomes. The chromosomes range in size over an order of magnitude, from the 1352 kbp chromosome IV to the 230 kbp chromosome I.

The yeast genome contains 5885 predicted protein-coding genes, ~140 genes for ribosomal RNAs, 40 genes for small nuclear RNAs, and 275 transfer RNA genes. In two respects, the yeast genome is denser in coding regions than the known genomes of the more complex eukàryotes Caenorhabditis elegans, Drosophila melanogaster, and human:

- Introns are relatively rare, and relatively small. Only 231 genes in yeast contain introns.

– There are fewer repeat sequences compared with more complex eukaryotes.

A duplication of the entire yeast genome appears to have occurred ~150 million years ago. This was followed by translocations of pieces of the duplicated DNA and loss of one of the copies of most (~92%) of the genes. Of the 5885 potential protein-coding genes, 3408 correspond to known proteins. About 1000 more contain some similarity to known proteins in other species.

Another ~800 are similar to ORFs in other genomes that correspond to unknown proteins. Many of these homologues appear in prokaryotes. Only approximately one-third of yeast proteins have identifiable homologues in the human genome. In taking censuses of genes, it has been useful to classify their functions into broad categories. The followings are classification of yeast protein functions:

– Metabolism
– Energy
– Cell growth, cell division and DNA synthesis
– Transcription
– Protein synthesis
– Protein destination
– Transport facilitation
– Cellular transport and transport mechanisms
– Cellular biogenesis
– Cellular communication/signal transduction
– Cell rescue, defence, cell death and ageing
– Ionic homeostasis
– Cellular organisation
– Transposable elements, viral and plasmid proteins
– Unclassified

Yeast is a testbed for development of methods to assign functions to gene products. The search for homologues has been exhaustive and continues. Collections of mutants exist containing a knockout of every gene. (A unique sequence 'bar code' introduced into each mutant facilitates identification of the ones that grow under selected conditions.)

A surprisingly-large fraction of these single-gene knockout strains are viable-80%. Cellular localisation and expression patterns are being investigated. Several types of measurements, including those based on activation of transcription by pairs of proteins that can form dimers-the two-hybrid method, are producing catalogues of interprotein interactions.

Genome of Caenorhabditis Elegans

The nematode worm Caenorhabditis elegans entered biological research in the 1960s, at the express invitation of S. Brenner. He recognised its potential as a sufficiently complex organism to be interesting but simple enough to permit complete analysis, at the cellular level, of its development and neural circuitry.

The C. elegans genome, completed in 1998, provided the first full DNA sequence of a multicellular organism. The C. elegans genome contains ~97 Mbp of DNA distributed on paired chromosomes I, II, III, IV, V and X. There is no Y chromosome. Different genders in C. elegans appear in the XX genotype, a selffertilising hermaphrodite; and the XO genotype, a male.The C. elegans genome is about 8 times larger than that of yeast, and its 19 099 predicted genes are approximately three times the number in yeast. The gene density is relatively low for a eukaryote, with ~1 gene/5 kbp of DNA. Exons cover ~27% of the genome; the genes contain an average of 5 introns. Approximately 25% of the genes are in clusters of related genes.

Many C. elegans proteins are common to other life forms. Others are apparently specific to nematodes. 42% of proteins have homologues outside the phylum, 34% are homologous to proteins of other nematodes and 24% have no known homologues outside C. elegans itself. Many of the proteins have been classified according to structure and function.

Several kinds of RNA genes have been identified. The C. elegans genome contains 659 genes for tRNA, almost half of them (44%) on the X chromosome. Spliceosomal RNAs appear in dispersed copies, often identical. (Spliceosomes are the organelles that convert pre-mRNA transcripts to mature mRNA by excising introns.) Ribosomal RNAs appear in a long tandem array at the end of chromosome I. 5S RNAs appear in a tandem array on chromosome V. Some RNA genes appear in introns of protein-coding genes.

The C. elegans genome contains many repeat sequences. Approximately 2.6% of the genome consists of tandem repeats. Approximately 3.6% of the genome contains inverted repeats; these appear preferentially within introns, rather than between genes. Repeats of the hexamer sequence TTAGGC appear in many places. There are also simple duplications, involving hundreds to tens of thousands of kbp.

Genome of Drosophila Melanogaster

Drosophila melanogaster, the fruit fly, has been the subject of detailed studies of genetics and development for a century. Its genome sequence, the product of a collaboration between Celera Genomics and the Berkeley Drosophila Genome Project, was announced in 1999.The chromosomes of D. melanogaster are nucleoprotein complexes. Approximately one-third of the genome is contained in heterochromatin, highly coiled and compact (and therefore densely staining) regions flanking the centromeres. The other two-thirds is euchromatin, a relatively uncoiled, less-compact form. Most of the active genes are in the euchromatin. The heterochromatin in D. melanogaster contains many tandem repeats of the sequence AATAACATAG, and relatively few genes. The total chromosomal DNA of Drosophila melanogaster contains about ~180 Mbp.

The released sequence consists of the euchromatic portion, about 120 Mbp. The genome is distributed over five chromosomes: three large autosomes, a Y chromosome, and a fifth tiny chromosome containing only ~1 Mbp of euchromatin. The fly's 13 601 genes are approximately double the number in yeast, but are fewer than in C. elegans, perhaps a surprise. The average density of genes in the euchromatin sequence is 1 gene/ 9 kbp; much lower than the typical 1gene/kbp densities of prokaryotic genomes.

Despite the fact that insects are not very closely related to mammals, the fly genome is useful in the study of human disease. It contains homologues of 289 human genes implicated in various diseases, including cancer, and cardiovascular, neurological, endocrinological, renal, metabolic, and haematological diseases. Some of these homologues have different functions in humans and flies. Other human disease-associated genes can be introduced into, and studied in, the fly.

For instance, the gene for human spinocerebellar ataxia type 3, when expressed in the fly, produces similar neuronal cell degeneration. There are now fly models for Parkinson disease and malaria. The noncoding regions of the D. melanogaster genome must contain regions controlling spatio-temporal patterns of development. The developmental biology of the fly has been studied very intensively. It is therefore an organism in which the study of the genomics of development will prove extremely informative.

Genome of Arabidopsis Thaliana

As a flowering plant Arabidopsis thaliana is a very distant relative of the other higher eukaryotic organisms, for which genome sequences are available. It invites comparative analysis to identify common and specialised features. The Arabidopsis thaliana genome contains ~125 Mbp of DNA, of which 115.4 Mbp was reported in 2000 by The Arabidopsis Genome Initiative. There are five pairs of chromosomes, containing 25 498 predicted genes. The genome is relatively compact, with 1 gene/4.6 kbp on average.

This figure is intermediate between prokaryotes and Drosophila, and roughly similar to C. elegans. The genes of Arabidopsis are relatively small. Exons are typically 250 bp long, and introns relatively small, with mean length 170 bp. Typical of plant genes is an enrichment of coding regions in GC content.

Most Arabidopsis proteins have homologues in animals, but some systems are unique, among higher organisms, to plants, including cell wall production and photosynthesis. Many proteins shared with animals have diverged widely since the last common ancestor. Typical of another difference between plants and animals, 25% of the nuclear genes have signal sequences governing their transport into organelles-mitochondria and chloroplasts-compared to 5% of mitochondrialtargeted nuclear genes in animals.

Traces of large- and small-scale duplication events appear in the Arabidopsis genome. 58% of the genome contains 24 duplicated segments, ≥ 100 kbp in length. The genes in the duplicated segments are usually not identifiable homologues. However, 4140 genes (17% of all genes) appear in 1528 tandem arrays of gene families, containing up to 23 members.

The pattern of duplications suggests that an ancestor of Arabidopsis underwent a whole-genome duplication, to form a tetraploid species. That this happened a long time-estimated at 112 million years-ago is shown by the very high amount of loss of one copy of duplicated genes, and wide divergence of those that remain. Higher plants must integrate the effects of three genomes-nuclear, chloroplast and mitochondrial. Many genes for proteins synthesised by nuclear genes and transported to organelles appear to have originated in the organelles and been transferred to the nucleus.

Genome of Homo Sapiens

In February 2001, the International Human Genome Sequencing Consortium and Celera Genomics published, separately, drafts of the human genome. The sequence amounted to $\sim 3.2 \times 10^9$ bp, thirty times larger than the genomes of C. elegans or D. melanogaster. One reason for this disparity in size is that coding sequences form less than 5% of the human genome; repeat sequences over 50%. Perhaps the most surprising feature was the small number of genes identified.

The finding of only about 20 000-25 000 genes suggests that alternative splicing patterns make a very significant contribution to our protein repertoire. It is estimated that ~50% of genes have alternative splicing patterns. The published human genome is distributed over 22 chromosome pairs plus the X and Y chromosomes. The DNA contents of the autosomes range from 279 Mbp down to 48 Mbp. The X chromosome contains 163 Mbp and the Y chromosome only 51 Mbp.

The exons of human protein-coding genes are relatively small compared to those in other known eukaryotic genomes. The introns are relatively long. As a result many protein-coding genes span long stretches of DNA. For instance, the dystrophin gene, coding for a 3685 amino acid protein, is >2.4 Mbp long.

Protein Coding Genes

Analysis of the human protein repertoire implied by the genome sequence has proved difficult because of the problems in reliably detecting genes, and alternative splicing patterns.

Repeat Sequences

Repeat sequences comprise over 50% of the genome:

- Transposable elements, or interspersed repeats-almost half the entire genome! These include the LINEs and SINEs.
- Retroposed pseudogenes
- Simple 'stutters'-repeats of short oligomers. These include the minisatellites and microsatellites. Trinucleotide repeats such as CAG, corresponding to glutamine repeats in the corresponding protein, are implicated in numerous diseases.
- Segmental duplications, of blocks of ~10-300 kbp. Interchromosomal duplications appear on nonhomologous chromosomes, sometimes at multiple sites. Some intrachromosomal duplications include closely spaced duplicated regions many kbp long of very similar sequence implicated in genetic diseases; for example Charcot-Marie-Tooth syndrome type 1A, a progressive peripheral neuropathy resulting from duplication of a region containing the gene for peripheral myelin protein 22.
- Blocks of tandem repeats, including gene families

SINGLE-NUCLEOTIDE POLYMORPHISMS (SNPS)

A single-nucleotide polymorphism or SNP (pronounced 'snip') is a genetic variation between individuals, limited to a single base pair which can be substituted, inserted or deleted. Sickle-cell anaemia is an example of a disease caused by a specific SNP: an A → T mutation in the β-globin gene changes a Glu → Val, creating a sticky surface on the haemoglobin molecule that leads to polymerisation of the deoxy form.

SNPs are distributed throughout the genome, occurring on average every 5000 base pairs. Although they arose by mutation, many positions containing SNPs have low mutation rates, and provide stable markers for mapping genes. A consortium of four major academic genome centres and eleven private companies is creating a high-quality, high-density human SNP map, with a commitment to open public access. A recently published version contains 1.8 million SNPs. Not all SNPs are linked to diseases.

Many are not within functional regions (although the density of SNPs is higher than the average in regions containing genes). Some SNPs that occur within exons are mutations to synonymous codons, or cause substitutions that do not significantly affect protein function. Other types of SNPs can cause more than local perturbation to a protein:

— A mutation from a sense codon to a stop codon, or vice versa, will cause either premature truncation of protein synthesis or 'readthrough'.

— A deletion or insertion will cause a phase shift in translation.

The A, B and O allelles of the genes for blood groups illustrate these possibilities. A and B alleles differ by four SNP substitutions. They code for related enzyme that add different saccharide units to an antigen on the surface of red blood cells.

Allele	Sequence	Saccharide
A	...gctggtgacccctt...	N-acetylgalactosamine
B	...gctcgtcaccgcta...	galactose
O	...cgtggt-acccctt...	—

The O allelle has undergone a mutation causing a phase shift, and produces no active enzyme. The red blood cells of type O individuals contain neither the A nor the B antigen. This is why people with type O blood are universal donors in blood transfusions. The loss of activity of the protein does not seem to carry any adverse consequences. Indeed, individuals of blood types B and O have greater resistance to smallpox.

Stong correlation of a disease with a specific SNP is advantageous in clinical work, because it is relatively easy to test for affected people or carriers. But if a disease arises from dysfunction of a specific protein, there ought to be many sites of mutations that could cause inactivation. A particular site may predominate if

— all bearers of the gene are descendants of a single individual in whom the mutation occurred, and/or

— the disease results from a gain rather than loss of a specific property, such as in the ability of sickle cell haemoglobin to polymerize, and/or

— the mutation rate at a particular site is unusually high, as in the Gly380'Arg mutation in the fibroblast growth receptor gene FGFR3, associated with achondroplasia (a syndrome including short stature).

In contrast, many independent mutations have been detected in the BRCA1 and BRCA2 genes, loci associated with increased disposition to early-onset breast and ovarian cancer. The normal gene products function as tumour suppressors. Insertion or deletion mutants causing phase shifts generally produce a missing or inactive protein. But it cannot be deduced a priori whether a novel substitution mutant in BRCA1 or BRCA2 confers increased risk or not. Treatments of diseases caused by defective or absent proteins include:

— Providing normal protein We have mentioned insulin for diabetes, and Factor VIII in the most common type of haemophilia. Another example is the administration of human growth hormone in patients with an absence or severe reduction in normal levels. Use of recombinant proteins eliminates the risk of transmission of AIDS through blood transfusions or of Creutzfeld-Jakob disease from growth hormone isolated from crude pituitary extracts.

– Lifestyle adjustments that make the function unnecessary Phenylketoneuria (PKU) is a genetic disease caused by deficiency in phenylalanine hydroxylase, the enzyme that converts phenylalanine to tyrosine. Accumulation of high levels of phenylalanine causes developmental defects, including mental retardation. The symptoms can be avoided by a phenylalanine-free diet. Screening of newborns for high blood phenylalanine levels is legally required in the United States and many other countries.

– Gene therapy to replace absent proteins is an active field of research. Other clinical applications of SNPs reflect correlations between genotype and reaction to therapy (pharmacogenomics). For example, a SNP in the gene for N-acetyl transferase (NAT-2) is correlated with peripheral neuropathy-weakness, numbness and pain in the arms, legs, hands or feet-as a side effect of treatment with isoniazid (isonicotinic acid hydrazide), a common treatment for tuberculosis. Patients who test positive for this SNP are given alternative treatment.

Genetic Diversity in Anthropology

SNP data are of great utility in anthropology, giving clues to historical variations in population size, and migration patterns. Degrees of genetic diversity are interpretable in terms of the size of the founding population. Founders are the original set of individuals from whom an entire population is descended. Founders can be either original colonists, such as the Polynesians who first settled New Zealand, or merely the survivors of a nearextinction.

Cheetahs show the effects of a population bottleneck, estimated to have occurred 10 000 years ago. All living cheetahs are as closely related to one another as typical siblings in other species. Extrapolations of mitochondrial DNA variation in contemporary humans suggest a single maternal ancestor who lived 14 000-200 000 years ago. Calling her Eve suggests that she was the first woman. But Mitochondrial Eve was in fact the founder of a surviving population following extinction of descendants of all her contemporaries.

Population-specific SNPs are informative about migrations. Mitochondrial sequences provide information about female ancestors and Y chromosome sequences provide information about male ancestors. For example, it has been suggested that the population of Iceland-first inhabited about 1100 years ago-is descended from Scandinavian males, and from females from both Scandinavia and the British Isles. Mediaeval Icelandic writings refer to raids on settlements in the British Isles.

A fascinating relationship between human DNA sequences and language families has been investigated by L. L. Cavalli-Sforza and colleagues. These studies have proved useful in working out interrelationships among American Indian languages. They confirm that the Basques, known to be a linguistically isolated population, have been genetically isolated also.

In the discovery of isolated populations, anthropological genetics provides data useful in medicine, for mapping disease genes is easier if the background variation is low. Genetically isolated populations in Europe include, in addition to the Basques, the Finns, Icelanders, Welsh, and Lapps.

Genetic Diversity and Personal Identification

Variations in our DNA sequences give us individual fingerprints, useful for identification and for establishment of relationships, including but not limited to questions of paternity. The use of DNA analysis as evidence in criminal trials is now well-established.

Genetic fingerprinting techniques were originally based on patterns of variable number tandem repeats (VNTRs), but have been extended to include analysis of other features including mitochondrial DNA sequences. For most of us, all our mitochondria are genetically identical, a condition called homoplasmy. However, some individuals contain mitochondria with different DNA sequences; this is heteroplasmy. Such sequence variation in a disease gene can complicate the observed inheritance pattern of the disease.

The most famous case of heteroplasmy involved Tsar Nicholas II of Russia. After the revolution in 1917 the Tsar and his family were taken to exile in Yekaterinburg in Central Russia. During the night of 16-17 July 1918, the Tsar, Tsarina Alexandra, at least three of their five children, plus their physician and three servants who had accompanied the family, were killed, and their bodies buried in a secret grave. When the remains were rediscovered, assembly of the bones and examination of the dental work suggested-and sequence analysis confirmed-that the remains included an expected family group.

The identity of the remains of the Tsarina were proved by matching the mitochondrial DNA sequence with that of a maternal relative, Prince Philip, Chancellor of the University of Cambridge, Duke of Edinburgh, and grandnephew of the Tsarina.

EVOLUTION OF GENOMES

The availability of complete information about genomic sequences has redirected research. A general challenge in analysis of genomes is to identify 'interesting events'. A background mutation rate in coding sequences is reflected in synonymous nucleotide substitutions: changes in codons that do not alter the amino acid. With this as a baseline one can search for instances in which there are significantly higher rates of nonsynonymous nucleotide substitutions: changes in codons that cause mutations in the corresponding protein.

Given two aligned gene sequences, we can calculate K_s = the number of synonymous substitutions, and K_a = the number of nonsynonymous substitutions. A high ratio of K_s/K_a identifies pairs of sequences apparently showing positive selection, possibly even functional changes.

The new field of comparative genomics treats questions such as:

- What genes do different phyla share? What genes are unique to different phyla? Do the arrangements of these genes in the genome vary from phylum to phylum?
- What homologous proteins do different phyla share? What proteins are unique to different phyla? Does the integration of the activities of these proteins vary from phylum to phylum? Do the mechanisms of control of expression patterns of these proteins vary from phylum to phylum?
- What biochemical functions do different phyla share? What biochemical functions are unique to different phyla? Does the integration of these biochemical functions vary from phylum to phylum? If two phyla share a function, and the protein that carries out this function in one phylum has a homologue in the other, does the homologous protein carry out the same function?

The same questions could be asked about different species within each phylum. Haemophilus influenzae represented the bacteria, Methanococcus jannaschii the archaea, and Saccharomyces cerevisiae (yeast) the eukarya. Their classification of protein functions contained as major categories processes involving energy, information, and communication and regulation:

General functional classes

- Energy
- Biosynthesis of cofactors, amino acids
- Central and intermediary metabolism
- Energy metabolism
- Fatty acids and phospholipids
- Nucleotide biosynthesis
- Transport
- Information
- Replication

- Transcription
- Translation
- Communication and regulation
- Regulatory functions
- Cell envelope/cell wall
- Cellular processes

The number of genes in the three species are:

Species	Number of genes
Haemophilus influenzae	1680
Methanococcus jannaschii	1735
Saccharomyces cerevisiae	6278

In the category of Energy, proteins are shared across the three domains. In the category of Communication and regulation, proteins are unique to each domain. In the Information category, archaea share some proteins with bacteria and others with eukarya.

Analysis of shared functions among all domains of life has led people to ask whether it might be possible to define a minimal organism-that is, an organism with the smallest gene complement consistent with independent life based on the central DNA → RNA → protein dogma. The minimal organism must have the ability to reproduce, but not be required to compete in growth and reproductive rate with other organisms.

Functional classes include:

- Translation, including protein synthesis
- DNA replication
- Recombination and repair-a second function of essential proteins involved in DNA replication.
- Transcription apparatus
- Chaperone-like proteins
- Intermediary metabolism-the glycolytic pathway
- No nucleotide, amino acid, or fatty acid biosynthesis
- Protein-export machinery
- Limited repertoire of metabolite transport proteins

It should be emphasised that the viability of an organism with these proteins has not been proven. Moreover, even if experiments proved that some minimal gene content-the proposed set or some other set-is necessary and sufficient, this does not answer the related question of identifying the gene complement of the common ancestor of M. genitalium and H. influenzae, or of the earliest cellular forms of life. For only 71% of the proposed set of 256 proteins have recognisable homologues among eukaryotic or archaeal proteins.

Nevertheless, identification of functions necessarily common to all forms of life allows us to investigate the extent to which different forms of life accomplish these functions in the same ways. Genome analysis has revealed families of proteins with homologues in archaea, bacteria, and eukarya. The assumption is that these have evolved from an individual ancestral gene through a series of speciation and duplication events, although some may be the effects of horizontal transfer.

The challenge is to map common functions and common proteins. Several thousand protein families have been identified with homologues in archaea, bacteria and eukarya. Different species contain different amounts of these common families: In bacteria, the range is from Aquifex aeolicus, 83% of the proteins of which have archaeal and eukaryotic homologues, to Borrelia burgdorferi, in which only 52% of the proteins have archaeal and eukáryotic homologues.

Archaeal genomes have somewhat higher percentages (62%-71%) of proteins with bacterial and eukaryotic homologues. But only 35% of the proteins of yeast have bacterial and archaeal homologues. Among the proteins of the minimal set identified from M. genitalium, only ~30% have homologues in all known genomes.

Horizontal Gene Transfer

In general, horizontal gene transfer is the acquisition of genetic material by one organism from another, by natural rather than laboratory procedures, through some means other than descent from a parent during replication or mating. Several mechanisms of horizonal gene transfer are known, including direct uptake, as in the pneumococcal transformation experiments, or via a viral carrier.

Analysis of genome sequences has shown that horizontal gene transfer is not a rare event, but has affected most genes in micro-organisms. It requires a change in our thinking from ordinary clonal or parental models of heredity. Evidence for horizontal transfer includes (1) discrepancies among evolutionary trees constructed from different genes, and (2) direct sequence comparisons between genes from different species:

- In E. coli, about 25% of the genes appear to have been acquired by transfer from other species.

— In microbial evolution, horizontal gene transfer is more prevalent among operational genes-those reponsible for 'housekeeping' activities such as biosynthesis-than among informational genes-those responsible for organisational activities such as transcription and translation. For example:

Bradyrhizobium japonicum, a nitrogen-fixing bacterium symbiotic with higher plants, has two glutamine synthetase genes. One is similar to those of its bacterial relatives; the other 50% identical to those of higher plants. Rubisco (ribulose-1,5-bisphosphate carboxylase/oxygenase), the enzyme that first fixes carbon dioxide at the entry to the Calvin cycle of photosynthesis, has been passed around between bacteria, mitochondria and algal plastids, as well as undergoing gene duplication.

Many phage genes appearing in the E. coli genome provide further examples and point to a mechanism of transfer. Nor is the phenomenon of horizontal gene transfer limited to prokaryotes. Both eukaryotes and prokaryotes are chimeras. Eukaryotes derive their informational genes primarily from an organism related to Methanococcus, and their operational genes primarily from proteobacteria with some contributions from cyanobacteria and methanogens. Almost all informational genes from Methanococcus itself are similar to those in yeast. At least 8 human genes appeared in the M. tuberculosis genome.

The observations hint at the model of a 'global organism', a genetic common market, or even a World-Wide DNA Web from which organisms download genes at will. How can this be reconciled with the fact that the discreteness of species has been maintained? The conventional explanation is that the living world contains ecological 'niches' to which individual species are adapted. It is the discreteness of niches that explains the discreteness of species. But this explanation depends on the stability of normal heredity to maintain the fitness of the species.

Perhaps the answer is that it is the informational genes, which appear to be less subject to horizontal transfer, that determine the identity of the species. It is interesting that although evidence for the importance of horizontal gene transfer is overwhelming, it was dismissed for a long time as rare and unimportant. The source of the intellectual discomfort is clear: Parent-to-child transmission of genes is at the heart of the Darwinian model of biological evolution whereby selection (differential reproduction) of parental phenotypes alters gene frequencies in the next generation.

For offspring to gain genes from elsewhere than their parents smacks of Lamarck and other discredited alternatives to the paradigm. The evolutionary tree as an organising principle of biological relationship is a deeply ingrained concept: scientists display an environmentalist-like fervour in their commitment to trees, even when trees are not an appropriate model of a network of relationships. Perhaps it is well to recall that Darwin knew nothing of genes, and the mechanism that generated the variation on which selection could operate was a mystery to him.

Comparative Genomics of Eukaryotes

A comparison of the genomes of yeast, fly, worm and human revealed 1308 groups of proteins that appear in all four. These form a conserved core of proteins for basic functions, including metabolism, DNA replication and repair, and translation.These proteins are made up of individual protein domains, including singledomain proteins, oligomeric proteins, and modular proteins containing many domains. The proteins of the worm and fly are built from a structural repertoire containing about three times as many domains as the proteins of yeast. Human proteins are built from about twice as many as those of the worm and fly.

Distribution of probable homologues of predicted human proteins

Vertebrates only	22%
Vertebrates and other animals	24%
Animals and other eukaryotes	32%
Eukaryotes and prokaryotes	21%
No homologues in animals	1%
Prokaryotes only	1%

To create new proteins, inventing new domains is a unusual event. It is far more common to create different combinations of existing domains in increasingly complex ways. A common mechanism is by accretion of domains at the ends of modular proteins. This process can occur independently, and take different courses, in different phyla.

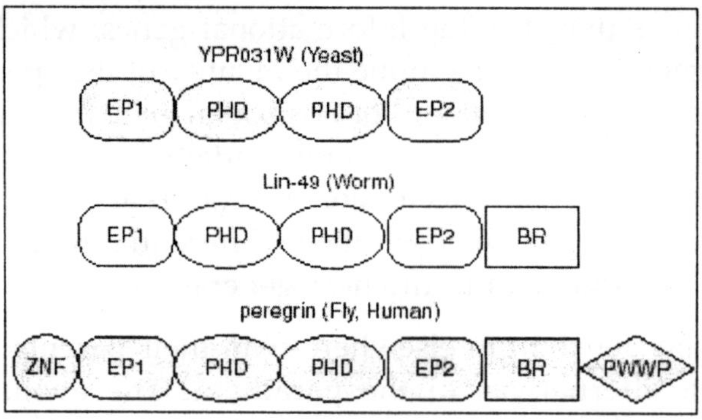

Figure 4. Evolution by accretion of domains, of molecules related to peregrin, a human protein that probably functions in transcription regulation.

Gene duplication followed by divergence is a mechanism for creating protein families. For instance, there are 906 genes + pseudogenes for olfactory receptors in the human genome. These are estimated to bind ~10 000 odour molecules. Homologues have been demonstrated in yeast and other fungi (some comparisons are odorous), but it is the need of vertebrates for a highly-developed sense of smell that multiplied and specialised the family to such a great extent. Eighty percent of the human olfactory receptor genes are in clusters. Compare the small size of the globin gene cluster, which did not require such great variety.

FULL GENOME SEQUENCING

Full genome sequencing (FGS), also known as whole genome sequencing, complete genome sequencing, or entire genome sequencing, is a laboratory process that determines the complete DNA sequence of an organism's genome at a single time. This entails sequencing all of an organism's chromosomal DNA as well as DNA contained in the mitochondria or chloroplast, depending respectively on whether the organism is an animal or plant. Almost any biological sample—even a very small amount of DNA or ancient DNA—can provide the genetic material necessary for full genome sequencing. Such samples may include saliva, epithelial cells, bone marrow, hair, seeds, plant leaves, or anything else that has DNA-containing cells. Because the sequence data that is produced can be quite large, genomic data is stored electronically and requires a large amount of computing power and storage capacity. Full genome sequencing would have been nearly impossible before the advent of the microprocessor, computers, and the Information Age.

Full genome sequencing should thus not be confused with DNA profiling. The latter only determines the likelihood that genetic material came from a particular individual or group and does not contain additional information on genetic relationships, origin or suspectability on specific diseases. It is also distinct from SNP genotyping which covers less than 0.1% of the genome. Almost all truly complete genomes are of microbes, the term "full genome" is sometimes used loosely to mean "greater than 95%". The remainder of this article focuses on nearly complete human genomes.

Full genome sequencing only refers to the laboratory process of deducing a person's entire genetic code and, on its own, may not contain any clinical assessment or useful clinical information. However, this may change over time as a large number of scientific studies continue to be published detailing clear associations between specific genetic variants and disease.

The first nearly complete human genomes sequenced were J. Craig Venter's and James Watson's, a Han Chinese (YH at 36-fold), a Yoruban from Nigeria (at 30-fold), a female leukemia patient, and Seong-Jin Kim. Other full genomes have been sequenced but not published, and as of June 2009, commercialization of full genome sequencing is in an early stage and growing rapidly.

One possible way to accomplish the cost-effective high-throughput sequencing necessary to accomplish full genome sequencing is by using Nanopore technology, which is a patented technology held by Harvard University and Oxford Nanopore Technologies and licensed to biotechnology companies. To facilitate their full genome sequencing initiatives, Illumina licensed nanopore sequencing technology from Oxford Nanopore Technologies and Sequenom licensed the technology from Harvard University. Another possible way to accomplish cost-effective high-throughput sequencing is by utilizing fluorophore technology. Pacific Biosciences is currently using this approach in their single molecule real time (SMRT) DNA sequencing technology. Complete Genomics is developing DNA Nanoball (DNB) technology that are arranged on self-assembling arrays. Pyrosequencing is a method of DNA sequencing based on the sequencing by synthesis principle. The technique was developed by Pål Nyrén and his student Mostafa Ronaghi at the Royal Institute of Technology in Stockholm in 1996, and is currently being used by 454 Life Sciences in their effort to deliver an affordable, fast and highly accurate full genome sequencing platform.

Sequencing of the entire human genome was first accomplished in 2000 partly through the use of shotgun sequencing technology. While full genome shotgun sequencing for small (4000–7000 base pair) genomes was already in use in 1979, broader application benefited from pairwise end sequencing, known colloquially as *double-barrel shotgun sequencing*. As sequencing projects began to take on longer and more complicated genomes, multiple groups began to realize that useful information could be obtained by sequencing both ends of a fragment of DNA. Although sequencing both ends of the same fragment and keeping track of the paired data was more cumbersome than sequencing a single end of two distinct fragments, the knowledge that the two sequences were oriented in opposite directions and were about the length of a fragment apart from each other was valuable in reconstructing the sequence of the original target fragment.

The first published description of the use of paired ends was in 1990 as part of the sequencing of the human HPRT locus, although the use of paired ends was limited to closing gaps after the application of a traditional shotgun sequencing approach. The first theoretical description of a pure pairwise end sequencing strategy, assuming fragments of constant length, was in 1991. In 1995 Roach et al. introduced the innovation of using fragments of varying sizes, and demonstrated that a pure pairwise end-sequencing strategy would be possible on large targets. The strategy was subsequently adopted by The Institute for Genomic Research (TIGR) to sequence the entire genome of the bacterium Haemophilus influenzae in 1995, and then by Celera Genomics to sequence the entire fruit fly genome in 2000, and subsequently the entire human genome. Applied Biosystems, now called Life Technologies, manufactured the shotgun sequencers utilized by both Celera Genomics and The Human Genome Project.

While shotgun sequencing was one of the first approaches utilized to successfully sequence the full genome of a human, it is too expensive and requires too long of a turn-around-time to be utilized for commercial purposes. Because of this, shotgun sequencing technology, even though it is still relatively 'new', is being displaced by technologies like pyrosequencing, SMRT sequencing, and nanopore technology.

In October 2006, the X Prize Foundation, working in collaboration with the J. Craig Venter Science Foundation, established the Archon X Prize for Genomics, intending to award US$10 million to "the first Team that can build a device and use it to sequence 100 human genomes within 10 days or less, with an accuracy of no more than one error in every 100,000 bases sequenced, with sequences accurately covering at least 98% of the genome, and at a recurring cost of no more than $10,000 per genome." However, higher accuracy rates are desirable for some clinical applications. An error rate of 1 in 100,000 bases, out of a total of six billion bases in the human diploid genome, would mean about 60,000 errors per genome, which is a significant number of false positives and negatives. For the latter it is not known where the errors occur . The error rates required for widespread clinical use, such as Predictive Medicine is currently set by over 1400 clinical single gene sequencing tests. As of June 2009, the Archon X Prize for Genomics remains unclaimed.

In 2007, Applied Biosystems started selling a new type of sequencer called SOLiD System, with the first sale to Helicos Biosciences in 2008. Helicos stated that, utilizing the new sequencers, they will attempt to provide a full genome sequencing service with a target price of $72,000 per sample. However, this price point is still too high some applications, and is only competitive to DNA arrays (at $500 per sample) in cases where more than 0.1% of the genome is desired.

In 2008 and 2009, both public and private companies have emerged that are now in a competitive race to be the first mover to provide a full genome sequencing platform that is commercially robust for both research and clinical use, including Illumina, Sequenom, 454 Life Sciences, Pacific Biosciences, Complete Genomics, Intelligent Bio-Systems, Genome Corp., and Helicos BioScience. These companies are heavily financed and backed by venture capitalists, hedge funds, investment banks and, in the case of Illumina, Sequenom and 454, heavy reinvestment of revenue into research and development, mergers and acquisitions, and licensing initiatives.

In the race to commercialize full genome sequencing, companies have made claims about being able to offer a service at a specific time for a specific price that have turned out to not be true. Intelligent Bio-Systems stated in November 2007 that by the end of 2008 they would release a platform capable of a providing a $5,000 full genome sequence, but, as of March 2009, no such platform has yet to be released.

Pacific Biosciences stated that they will start selling their full genome sequencers in early 2010. While they didn't disclose the cost to sequence a single genome, they did state they may not release their second-generation machine capable of a $1,000 genome until 2013. Complete Genomics, however, stated that they'll be able to provide a $5,000 full genome sequencing service by the summer of 2009. The accuracy, prècision, and reproducibility of both Pacific Biosciences and Complete Genomics technology, however, is still unknown.

A personal genomics company located in Massachusetts, currently provides genome sequencing services but the cost is about $99,500 per genome, the turn-around time is unknown, the accuracy is unknown, and the number of people was limited to 20 for the first year, and is still considered early stage commercialization of full genome sequencing, focusing on wealthy customers. As of January 2009, there are no indications that any of these companies have been hindered by the global recession. And thus, the race appears to be proceeding forward at full speed.

At the end of February 2009, Complete Genomics released a full sequence of a human genome that was sequenced using their service. The data indicates that Complete Genomics' full genome sequencing service accuracy is just under 99.99%, meaning that there is an error in one out of every ten thousand base pairs. This means that their full sequence of the human genome will contain approximately 80,000-100,000 false positive errors in each genome. However, this accuracy rate was based on Complete Genomics' sequence that was completed utilizing a 90x depth of coverage while their commercialized sequence is reported to be only 40x, so the accuracy may be substantially lower unless they can find some way to improve it before their first service release planned for the summer 2009. This accuracy rate may be acceptable for research purposes, and clinical use would require confirmation by other methods of any reportable alleles. In March 2009, it was announced that Complete Genomics has signed a deal with the Broad Institute to sequence cancer patient's genomes and will be sequencing five full genomes to start. In April 2009, Complete Genomics announced that it plans to sequence 1,000 full genome's between June 2009 and the end of the year and that they plan to be able to sequence one million full genomes *per year* by 2013. Complete Genomics plans to officially launch in June 2009, although it is unknown if their lab will have received CLIA-certification by that time. In June 2009, Illumina announced that they were launching their own Personal Full Genome Sequencing Service at a depth of 30X for $48,000 per genome. This is still expensive for widespread consumer use, but the price may decrease substantially over the next few years as they realize economies of scale and given the competition with other companies such as Complete Genomics. Jay Flatley, Illumina's President & CEO, stated that "during the next five years, perhaps markedly sooner," the price point for full genome sequencing will fall from $48,000 to under $1,000. Illumina has already signed agreements to supply full genome sequencing services to multiple direct-to-consumer personal genomics companies.

Disruptive Technology

Full genome sequencing provides information on a genome that is orders of magnitude larger than that provided by the current leader in sequencing technology, DNA arrays.

For humans, DNA arrays currently provides genotypic information on up to one million genetic variants, while full genome sequencing will provide information on all six billion bases in the human genome, or 3,000 times more data. Because of this, full genome sequencing is considered disruptive to the DNA array markets as the accuracy of both range from 99.98% to 99.999% and their cost of $5000 per 6 billion base pairs is competitive with DNA arrays ($500 per 1 million basepairs).

Agilent, another established DNA array manufacturer, is working on targeted genome sequencing technologies. It is thought that Affymetrix, the pioneer of array technology in the 1990s, has fallen behind due to significant corporate and stock turbulence and is currently not working on any known full genome sequencing approach. It is unknown what will happen to the DNA array market once full genome sequencing becomes commercially widespread, especially as companies and laboratories providing this disruptive technology start to realize economies of scale. It is postulated, however, that this new technology may significantly diminish the total market size for arrays and any other sequencing technology once it becomes commonplace for individuals and newborns to have their full genomes sequenced.

Societal Impact

Inexpensive, time-efficient full genome sequencing will be a major accomplishment not only for the field of Genomics, but for the entire human civilization because, for the first time, individuals will be able to have their entire genome sequenced. Utilizing this information, it is speculated that health care professionals, such as physicians and genetic counsellors, will eventually be able to use genomic information to predict what diseases a person may get in the future and attempt to either minimize the impact of that disease or avoid it altogether through the implementation of personalized, preventive medicine. Full genome sequencing will allow health care professionals to analyse the entire human genome of an individual and therefore detect all disease-related genetic variants, regardless of the genetic variant's prevalence or frequency. This will enable the rapidly emerging medical fields of Predictive Medicine and Personalized Medicine and will mark a significant leap forward for the clinical genetic revolution.

Full genome sequencing is clearly of great importance for research into the basis of genetic disease. However, it should be recognized that despite advancements in genome sequencing technology, incomplete understanding of the significance of individual variants or combinations of variants will limit the widespread usefulness of full genome sequencing in medicine until its clinical utility can be demonstrated.

DNA MICROARRAY

A DNA microarray is a multiplex technology used in molecular biology and in medicine. It consists of an arrayed series of thousands of microscopic spots of DNA oligonucleotides, called features, each containing picomoles of a specific DNA sequence. This can be a short section of a gene or other DNA element that are used as probes to hybridise a cDNA or cRNA sample under high-stringency conditions. Probe-target hybridisation is usually detected and quantified by detection of fluorophore-, silver-, or chemiluminescence-labeled targets to determine relative abundance of nucleic acid sequences in the target.

In standard microarrays, the probes are attached to a solid surface by a covalent bond to a chemical matrix. The solid surface can be glass or a silicon chip, in which case they are commonly known as *gene chip* or colloquially *Affy chip* when an Affymetrix chip is used. Other microarray platforms, such as Illumina, use microscopic beads, instead of the large solid support. DNA arrays are different from other types of microarray only in that they either measure DNA or use DNA as part of its detection system.

DNA microarrays can be used to measure changes in expression levels, to detect single nucleotide polymorphisms (SNPs), in genotyping or in resequencing mutant genomes. Microarrays also differ in fabrication, workings, accuracy, efficiency, and cost. Additional factors for microarray experiments are the experimental design and the methods of analysing the data.

Microarray technology evolved from Southern blotting, where fragmented DNA is attached to a substrate and then probed with a known gene or fragment. The use of a collection of distinct DNAs in arrays for expression profiling was first described in 1987, and the arrayed DNAs were used to identify genes whose expression is modulated by interferon. These early gene arrays were made by spotting cDNAs onto filter paper with a pin-spotting device. The use of miniaturised microarrays for gene expression profiling was first reported in 1995, and a complete eukaryotic genome (*Saccharomyces cerevisiae*) on a microarray was published in 1997.

Arrays of DNA can be spatially arranged, as in the commonly known *gene chip*, or can be specific DNA sequences labelled such that they can be independently identified in solution. The traditional solid-phase array is a collection of microscopic DNA spots attached to a solid surface, such as glass, plastic or silicon biochip. The affixed DNA segments are known as *probes*. Thousands of them can be placed in known locations on a single DNA microarray.

DNA microarrays can be used to detect DNA, or detect RNA that may or may not be translated into proteins. The process of measuring gene expression via cDNA is called expression analysis or expression profiling. Since an array can contain tens of thousands of probes, a microarray experiment can accomplish that many genetic tests in parallel. Therefore arrays have dramatically accelerated many types of investigation. Applications include:

Technology or Application	*Synopsis*
Gene expression profiling	In an mRNA or gene expression profiling experiment the expression levels of thousands of genes are simultaneously monitored to study the effects of certain treatments, diseases, and developmental stages on gene expression. For example, microarray-based gene expression profiling can be used to identify genes whose expression is changed in response to pathogens or other organisms by comparing gene expression in infected to that in uninfected cells or tissues.
Comparative genomic hybridisation	Assessing genome content in different cells or closely related organisms.
GeneID	Small microarrays to check IDs of organisms in food and feed, mycoplasms in cell culture, or pathogens for disease detection, mostly combining PCR and microarray technology.
Chromatin immunoprecipitation on Chip	DNA sequences bound to a particular protein can be isolated by immunoprecipitating that protein (ChIP), these fragments can be then hybridised to a microarray allowing the determination of protein binding site occupancy throughout the genome. Example protein to immunoprecipitate are histone modifications, Polycomb-group protein and trithorax-group protein to study the epigenetic landscape or RNA Polymerase II to study the transcription landscape.
SNP detection	Identifying single nucleotide polymorphism among alleles within or between populations. Several applications of microarrays make use of SNP detection, including Genotyping, forensic analysis, measuring predisposition to disease, identifying drug-candidates, evaluating germline mutations in individuals or somatic mutations in cancers, assessing loss of heterozygosity, or genetic linkage analysis.
Alternative splicing detection	An *'exon junction array* design uses probes specific to the expected or potential splice sites of predicted exons for a gene. It is of intermediate density, or coverage, to a typical gene expression array (with 1-3 probes per gene) and a genomic tiling array. It is used to assay the expression of alternative splice forms of a gene.
Fusion genes microarray	A Fusion gene microarray can detect fusion transcripts, e.g. from cancer specimens. The principle behind this is building on the alternative splicing micorrays. The oligo design strategy enables combined measurements of chimeric transcript junctions with exon-wise measurements of individual fusion partners.
Tiling array	Genome tiling arrays consist of overlapping probes designed to densely represent a genomic region of interest, sometimes as large as an entire human chromosome. The purpose is to empirically detect expression of transcripts or alternatively splice forms which may not have been previously known or predicted.

Fabrication

Microarrays can be manufactured in different ways, depending on the number of probes under examination, costs, customisation requirements, and the type of scientific question being asked. Arrays may have as few as 10 probes or up to 2.1 million micrometre-scale probes from commercial vendors.

Surface engineering

The first step of DNA microarray fabrication involves surface engineering of a substrate in order to obtain desirable surface properties for the application of interest. Optimal surface properties are those which produce high signal to noise ratios for the DNA targets of interest. Generally, this involves maximising the probe surface density and activity while minimising the non-specific binding of the targets of interest. Methods of surface engineering vary depending on the platform material, design, and application.

Spotted vs. oligonucleotide arrays

Microarrays can be fabricated using a variety of technologies, including printing with fine-pointed pins onto glass slides, photolithography using pre-made masks, photolithography using dynamic micromirror devices, ink-jet printing, or electrochemistry on microelectrode arrays.

In spotted microarrays, the probes are oligonucleotides, cDNA or small fragments of PCR products that correspond to mRNAs. The probes are synthesized prior to deposition on the array surface and are then "spotted" onto glass. A common approach utilises an array of fine pins or needles controlled by a robotic arm that is dipped into wells containing DNA probes and then depositing each probe at designated locations on the array surface. The resulting "grid" of probes represents the nucleic acid profiles of the prepared probes and is ready to receive complementary cDNA or cRNA "targets" derived from experimental or clinical samples. This technique is used by research scientists around the world to produce "in-house" printed microarrays from their own labs.

These arrays may be easily customised for each experiment, because researchers can choose the probes and printing locations on the arrays, synthesize the probes in their own lab, and spot the arrays. They can then generate their own labeled samples for hybridisation, hybridise the samples to the array, and finally scan the arrays with their own equipment. This provides a relatively low-cost microarray that may be customised for each study, and avoids the costs of purchasing often more expensive commercial arrays that may represent vast numbers of genes that are not of interest to the investigator. Publications exist which indicate in-house spotted microarrays may not provide the same level of sensitivity compared to commercial oligonucleotide arrays, possibly owing to the small batch sizes and reduced printing efficiencies when compared to industrial manufactures of oligo arrays.

In oligonucleotide microarrays, the probes are short sequences designed to match parts of the sequence of known or predicted open reading frames. Although oligonucleotide probes are often used in "spotted" microarrays, the term "oligonucleotide array" most often refers to a specific technique of manufacturing. Oligonucleotide arrays are produced by printing short oligonucleotide sequences designed to represent a single gene or family of gene splice-variants by synthesizing this sequence directly onto the array surface instead of depositing intact sequences.

Sequences may be longer or shorter depending on the desired purpose; longer probes are more specific to individual target genes, shorter probes may be spotted in higher density across the array and are cheaper to manufacture. One technique used to produce oligonucleotide arrays include photolithographic synthesis on a silica substrate where light and light-sensitive masking agents are used to "build" a sequence one nucleotide at a time across the entire array. Each applicable probe is selectively "unmasked" prior to bathing the array in a solution of a single nucleotide, then a masking reaction takes place and the next set of probes are unmasked in preparation for a different nucleotide exposure. After many repetitions, the sequences of every probe become fully constructed. More recently, Maskless Array Synthesis from NimbleGen Systems has combined flexibility with large numbers of probes.

Two-channel vs. one-channel detection

Two-color microarrays or two-channel microarrays are typically hybridised with cDNA prepared from two samples to be compared and that are labeled with two different fluorophores. Fluorescent dyes commonly used for cDNA labelling include Cy3, which has a fluorescence emission wavelength of 570 nm, and Cy5 with a fluorescence emission wavelength of 670 nm. The two Cy-labelled cDNA samples are mixed and hybridised to a single microarray that is then scanned in a microarray scanner to visualise fluorescence of the two fluorophores after excitation with a laser beam of a defined wavelength. Relative intensities of each fluorophore may then be used in ratio-based analysis to identify up-regulated and down-regulated genes.

Oligonucleotide microarrays often contain control probes designed to hybridise with RNA spike-ins. The degree of hybridisation between the spike-ins and the control probes is used to normalise the hybridisation measurements for the target probes. Although absolute levels of gene expression may be determined in the two-color array, the relative differences in expression among different spots within a sample and between samples is the preferred method of data analysis for the two-color system. Examples of providers for such microarrays includes Agilent with their Dual-Mode platform, Eppendorf with their DualChip platform for colorimetric Silverquant labeling, and TeleChem International with Arrayit.

In *single-channel microarrays* or *one-color microarrays*, the arrays are designed to give estimations of the absolute levels of gene expression. Therefore the comparison of two conditions requires two separate single-dye hybridisations. As only a single dye is used, the data collected represent absolute values of gene expression. These may be compared to other genes within a sample or to reference "normalising" probes used to calibrate data across the entire array and across multiple arrays.

Three popular single-channel systems are the Affymetrix "Gene Chip", the Applied Microarrays "CodeLink" arrays, and the Eppendorf "DualChip & Silverquant". One strength of the single-dye system lies in the fact that an aberrant sample cannot affect the raw data derived from other samples, because each array chip is exposed to only one sample. Another benefit is that data are more easily compared to arrays from different experiments; the absolute values of gene expression may be compared between studies conducted months or years apart. A drawback to the one-color system is that, when compared to the two-color system, twice as many microarrays are needed to compare samples within an experiment.

Microarrays and Bioinformatics

The advent of inexpensive microarray experiments created several specific bioinformatics challenges:

- the multiple levels of replication in experimental design (Experimental design)
- the number of platforms and independent groups and data format (Standardisation)
- the treatment of the data (Statistical analysis)
- accuracy and precision (Relation between probe and gene)
- the sheer volume of data and the ability to share it (Data warehousing)

Experimental design

Due to the biological complexity of gene expression, the considerations of experimental design that are discussed in the expression profiling article are of critical importance if statistically and biologically valid conclusions are to be drawn from the data.

There are three main elements to consider when designing a microarray experiment. First, replication of the biological samples is essential for drawing conclusions from the experiment. Second, technical replicates help to ensure precision and allow for testing differences within treatment groups. The technical replicates may be two independent RNA extractions or two aliquots of the same extraction. Third, spots of each cDNA clone or oligonucleotide are present as replicates on the microarray slide, to provide a measure of technical precision in each hybridisation. It is critical that information about the sample preparation and handling is discussed, in order to help identify the independent units in the experiment and to avoid inflated estimates of statistical significance.

Standardisation

Microarray data is difficult to exchange due to the lack of standardisation in platform fabrication, assay protocols, and analysis methods. This presents an interoperability problem in bioinformatics. Various grass-roots open-source projects are trying to ease the exchange and analysis of data produced with non-proprietary chips:

- For example, the "Minimum Information About a Microarray Experiment" (MIAME) checklist helps define the level of detail that should exist and is being adopted by many journals as a requirement for the submission of papers incorporating microarray results. But MIAME does not describe the format for the information, so while many formats can support the MIAME requirements, as of 2007 no format permits verification of complete semantic compliance.
- The "MicroArray Quality Control (MAQC) Project" is being conducted by the US Food and Drug Administration (FDA) to develop standards and quality control metrics which will eventually allow the use of MicroArray data in drug discovery, clinical practice and regulatory decision-making.
- The MGED Society has developed standards for the representation of gene expression experiment results and relevant annotations.

Statistical analysis

The analysis of DNA microarrays poses a large number of statistical problems, including the normalisation of the data. There are several normalisation methods in the published literature some of which are platform specific; as in many other cases where authorities disagree, a sound conservative approach is to directly compare different normalisation methods to determine the effects of these different methods on the results obtained. This can be done, for example, by investigating the performance of various methods on data from "spike-in" experiments.

Also, experimenters must account for multiple comparisons: even if the statistical P-value assigned to a gene indicates that it is extremely unlikely that differential expression of this gene was due to random rather than treatment effects, the very high number of genes on an array makes it likely that differential expression of some genes represent false positives or false negatives. Statistical methods tailored to microarray analyses have recently become available that assess statistical power based on the variation present in the data and the number of experimental replicates, and can help minimise type I and type II errors in the analyses.

A basic difference between microarray data analysis and much traditional biomedical research is the dimensionality of the data. A large clinical study might collect 100 data items per patient for thousands of patients. A medium-size microarray study will obtain many thousands of numbers per sample for perhaps a hundred samples. Many analysis techniques treat each sample as a single point in a space with thousands of dimensions, then attempt by various techniques to reduce the dimensionality of the data to something humans can visualise. An example for such a method is the Local Pooled Error (LPE) test, which pools standard deviations of genes with similar expression levels and thereby overcomes the problem of low replicate numbers.

Relation between probe and gene

The relation between a probe and the mRNA that it is expected to detect is problematic. On the one hand, some mRNAs may cross-hybridise probes in the array that are supposed to detect another mRNA. On the other hand, probes that are designed to detect the mRNA of a particular gene may be relying on genomic EST information that is incorrectly associated with that gene.

REFERENCES

Bloomfield,V.A., Crothers,D.M., and Tinoco, I., Jr., *Nucleic Acids. Structures, Properties, and Functions,* University Science Books, 2000.

Bongaerts, J., Krämer, M., Müller, U., Raeven, 1.. and Wubbolts, M., "Metabolic engineering for microbial production of aromatic amino acids and derived compounds", *Metabolic* Engineering, 2001.

Dobson CM., "The nature and significance of protein folding", In *Mechanisms of Protein Folding* 2nd ed. Ed. RH Pain, *Frontiers in Molecular Biology series,* Oxford University Press: New York, NY, 2000.

Dickerson, R.E., "DNA structure from A to Z", *Methods Enzymol,* 1992.

Hodgson, J., "Bulk amino acid fermentation: technology and commodity trading", *Bio/Technology,* 1994.

Ikeda, M., "Amino acid production processes", *Advances Biochernical Engineering,* 2003.

Kuhlman B, Dantas G, Ireton G.C, Varani G, Stoddard B.L, Baker D., "Design of a novel globular protein fold with atomic-level accuracy", *Science,* 2003.

Ohnishi, J., Mitsuhashi, S., Llayashi, M., et al., "A novel methodology employing *Corynebacterium glutamicum* genome information to generate a new r-lysineproducing mutant", *Appliecl Microbiology and Biotechnology,* 2002.

Thieffry, D., "Forty years under the central dogma", *Trends Biochem. Sci.,* 1998.

Watson, J.D. and Crick, F.H.C., "Molecular structure of nucleic acids", *Nature,* 1953.

Walian P, Cross T.A, Jap B.K., "Structural genomics of membrane proteins", *Genome Biology,* 2004.

4

Molecular Modelling

Technology today lacks a fundamental capability: we are unable to build complex systems to atomic specifications. This typically limits not only what we build, but what we think about. Engineers attempt to fit their designs to the limits of today's materials and fabrication techniques, for obvious reasons–there is little profit or strategic advantage in a design that cannot be built. But will build tomorrow more than we can build today. New tools will let us build better systems, including yet better tools. And it is not futile to try to understand what may be built with tomorrow's tools: the laws of nature do not change as technology does, and many laws seem well understood. With today's science and a willingness to think about tomorrow's tools, it may be that we can see the shape of tomorrow's engineering. A key development will be the molecular assembler.

Assemblers will be molecular machines that direct chemical bonding processes through the manipulation of reactive molecules. They will make possibile *nanotechnology*, technology based on the ability to build systems to complex, atomic specifications. As microtechnology operates on a micrometer scale, so nanotechnology will operate on a nanometer scale. This definition of nanotechnology refers to more than just scale, though–the ability to etch lines a nanometer wide does not imply the ability to build molecular machines and computers.

The idea of molecular machinery was latent in the earliest descriptions of molecules as mechanical systems. The kinetic theory of gases marked an early milestone; the recognition that biochemical processes rest on molecular machines marked another. Discussion of molecular assemblers and a technology of molecular machines, however, is apparently quite recent.

Richard Feynman's 1959 talk on miniaturization foreshadowed the concept of the assembler, though it lacked any concrete discussion of chemical bonding or molecular machinery. Feynman observed that "The principles of physics, as far as I can see, do not speak against the possibility of maneuvering things atom by atom," and went on to remark: "But it is interesting that it would be, in principle, possible (I think) for a physicist to synthesize any chemical substance that the chemist writes down. Give the orders and the physicist synthesizes it. How? Put the atoms down where the chemist says, and so you make the substance." Thus, while Feynman did not describe the assembler, he was clearly looking in a similar direction.

The case for the assembler can be made most briefly through consideration of existing biochemical machinery. Enzymes, ribosomes, and the genetic system demonstrate that molecular machines can exist, that they can position and manipulate reactive molecules, and that they can guide the breakage and formation of chemical bonds in specific ways under programmable control. These biochemical machines show lower bounds to the capabilities of general molecular machines; they also indicate that protein engineering offers a path to the development of novel molecular machinery.

The greatest limit to modern synthetic chemistry is its limited control of reaction sites. The use of even primitive assemblers to position reactive molecules will give far more flexible control of where chemical reactions occur on large molecules: this will permit the creation of molecular structures far more complex than any that synthetic chemists can now construct. Today chemists mix molecules in solution, letting them tumble randomly and react wherever they can; achieving site specificity becomes increasingly difficult as the size of the molecules–and hence the number of possible reaction sites–grows large. Assemblers will place chemists in a fundamentally different situation. An assembler will resemble an industrial robot arm, able to move a part to a specific position with respect to a workpiece. This positioning ability will let chemists make even the most indiscriminately reactive molecules bond specifically, to build up specific molecular structures.

Among these structures will be components for molecular machines. Thus, however the first assembler may be built, we can expect to see an iterative process in which assemblers are used to build better assemblers. The use of strong, rigid materials in assemblers will aid the application of mechanical energy to chemical reactions; components with suitable electrical properties will do likewise for the use of electrical energy. Together, these advances will make possible the synthesis of an extremely broad class of structures, which may be expected to include virtually all molecular structures of engineering interest. The protein-engineering path to assemblers is already being followed. Other paths seem possible: these include exploitation of non-protein supramolecular chemistry and the development of manipulators based on the positioning technology used in scanning tunneling microscopess. For understanding the capabilities of future technology, however, the path to be followed is unimportant:

All paths lead ultimately to assemblers able to build systems to complex, atomic specifications, including systems built from independently assembled parts. It is thus of interest to consider what sorts of systems can be built, given the assumption that essentially any well-bonded molecular structure can be synthesized, and that these structures can be combined to build complex systems.

MOLECULAR ENGINEERING

Molecular engineering is any means of manufacturing molecules. It may be used to create, on an extremely small scale, most typically one at a time, new molecules which may not exist in nature, or be stable beyond a very narrow range of conditions.Today this is an extremely difficult process, requiring manual manipulation of molecules using such devices as a scanning tunneling microscope. Eventually it is expected to exploit life-like self-replicating 'helper molecules' that are themselves engineered. Thus the field can be seen as a precision form of chemical engineering that includes protein engineering, the creation of protein molecules, a process that occurs naturally in biochemistry, e.g., prion reproduction. However, it provides far more control than genetic modification of an existing genome, which must rely strictly on existing biochemistry to express genes as proteins, and has little power to produce any non-proteins.

Molecular engineering is an important part of pharmaceutical research and materials science. Emergence of scanning tunneling microscopes and picosecond-burst lasers in the 1990s, plus discovery of new carbon nanotube applications to motivate mass production of these custom molecules, drove the field forward to commercial reality in the 2000s.As it matures, it is seeming to converge with mechanical engineering, since the molecules being designed often resemble small machines. A general theory of molecular mechanosynthesis to parallel that of photosynthesis and chemosynthesis is the ultimate goal of the field. This may lead to a molecular assembler, according to some, such as K. Eric Drexler, Ralph Merkle, and Robert Freitas, and of the potential for integrating vast numbers of assemblers into a kg-scale nanofactory.Molecular engineering is sometimes called generically "nanotechnology", in reference to the nanometre scale at which its basic processes must operate. That term is considered to be vague, however, due to misappropriation of the word in association with other techniques, such as X-ray lithography, that are not used to create new free-floating ions or molecules.

MOLECULAR MACHINES

For engineers thinking about molecular machines, it is important to recognize the extent to which physical intuition and familiar mechanical principles still apply. Although describing the motion of electrons on a molecular scale demands the use of quantum mechanics, describing the motion of molecules as a whole generally does not. Newtonian mechanics is adequate for most purposes, and forms the basis of an extensive body of work in the field of computational molecular mechanics. Molecules have well-defined sizes and shapes (though they have soft surfaces). Molecular bonds act much like structural members, with preferred directions and lengths set by elastic restoring forces. What is more, these intra- and inter-molecular forces are familiar to everyone–they are the same forces that hold together larger collections of atoms, such as tables and aircraft.

In short, molecules are simply the smallest instances of material objects. Molecules do, of course, have unique properties from an engineering perspective. Atoms are like prefabricated parts, but atoms of a given type are utterly identical and not subject to wear. Machines made by bonding together these parts are utterly identical, if they are built with the same pattern of atoms and bonds. Accordingly, they show no wear or fatigue, no subtle changes over time: they are either whole, or broken.

Molecular machines suffer from unusual problems and failure mechanisms. Thermal noise has significant effects, making parts shake to an extent that varies inversely with their stiffness. Slim parts many atoms long suffer vibrations that become significant on an atomic scale. Radiation damage is a major concern: it causes bond breakage and device failure at a rate that gives hundred-million atom systems a half-life of a few centuries under ambient terrestrial conditions8. This half-life for damage to occur is inversely proportional to system volume. Accordingly, large systems of molecular machinery will require redundant, fault-tolerant design.

The scaling laws for mechanical systems are such that the frequencies associated with moving parts are inversely proportional to their linear dimensions, all else being equal. Scaling industrial robot arms down to the 0.1 micron scale of an assembler arm gives a ten million-fold increase in operations per second. Mechanical computers built with parts of subnanometer scale will have clock frequencies in the gigahertz range.

Biological analogies suggest two means of powering molecular machinery: chemical energy and light, the energy sources of animal and plant cells. Design work indicates that electric motors can be scaled down to a size of about 50 nanometers, making electrical power another alternative.

REPLICATING ASSEMBLERS

Assemblers will be able to build virtually any specified arrangement of atoms consistent with the constraints of chemical bónding. Since they themselves will be instances of arrangements of atoms, properly programmed assemblers will be able to build copies of themselves. Compared to proposals for self-replicating machines based on conventional technology, this approach has several advantages. First, the greater frequency of operation (the frequency of arm motions can be in the megahertz range) will allow replicating assemblers to have short generation times, estimated to be in the range of hundreds to thousands of seconds (comparable to those of existing molecular systems, such as bacteria). Second, the use of atoms as prefabricated parts eliminates the parts-manufacture problem: imagine designing a self-replicating machine to operate in a warehouse full of quality-controlled parts suited for just such a machine. Finally, replicating assemblers will be working with the ultimate building blocks of matter, allowing them to be reprogrammed after replication to build the widest possible range of products.

We are already familiar with self-replicating molecular machines: crabgrass is an example. These systems show that molecular replicators can produce products of incredible complexity in vast quantity and at negligible cost. Once the formidable challenges of tooling and design have been surmounted, replicating assemblers can do the same for products designed by engineers.

ASSEMBLER-BUILT PRODUCTS

Assemblers lend themselves naturally to the construction of small structures, such as devices for computers. Thus, they lend themselves to the implementation of molecular electronic systems. Comparatively detailed design work on molecular mechanical computers indicates that the equivalent of a modern mainframe computer can fit in less than a cubic micron, while consuming on the order of 10 nanowatts of power. This has implications both for lightweight, low-power systems, and for future massively parallel computers.

Large numbers of assemblers in coordination lend themselves to the construction of large structures made of nearly flawless materials. Properly arranged carbon atoms will make diamond fiber, which can serve as the basis for tough, atomically-tailored fiber composites. Diamond has a tensile strength of about 50 gigapascal, a modulus of about 1000 gigapascal, and a specific gravity of about 3.5. Again, once the formidable challenges of tooling and design have been surmounted, it seems that hardware based on these materials can become inexpensive–indeed, there seems no reason why such hardware should cost more than unmilled wood does today, since both will be carbon-rich products of self-replicating molecular machinery.

If these "large structures made of nearly flawless materials" are spacecraft, then space travel will become inexpensive. Strong materials will make possible high-performance vehicles, while simultaneously allowing broad margins of safety to increase reliability and reduce maintenance requirements. The energy cost of access to space is already low, and solar collectors (for example) built by replicating assemblers can make energy costs lower yet. When reliable, low-maintenance, high-performance vehicles finally become readily available, the main costs of spaceflight will have disappeared.

Molecular mechanics can be used to study small molecules as well as large biological systems or material assemblies with many thousands to millions of atoms.

All-atomistic molecular mechanics methods have the following properties:

- Each atom is simulated as a single particle
- Each particle is assigned a radius (typically the van der Waals radius), polarizability, and a constant net charge (generally derived from quantum calculations and/or experiment)

— Bonded interactions are treated as "springs" with an equilibrium distance equal to the experimental or calculated bond length

Variations on this theme are possible; for example, many simulations have historically used a "united-atom" representation in which methyl and methylene groups were represented as a single particle, and large protein systems are commonly simulated using a "bead" model that assigns two to four particles per amino acid.

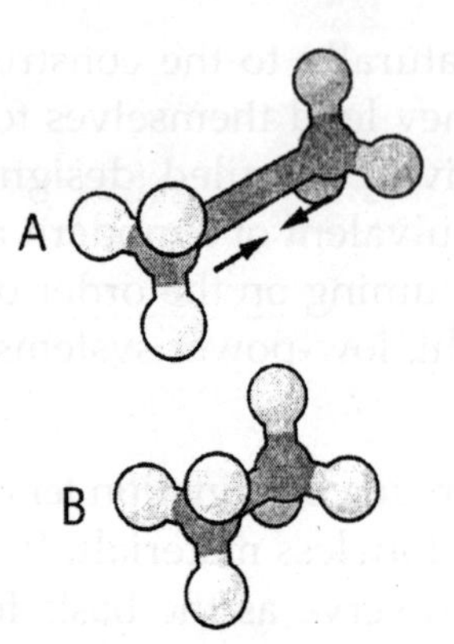

Figure 1. A force field is used to minimize the bond stretching energy of this ethane molecule

The following functional abstraction, known as a potential function or force field in Chemistry, calculates the molecular system's potential energy (E) in a given conformation as a sum of individual energy terms.

$$E = E_{covalent} + E_{noncovalent}$$

$$E_{covalent} = E_{bond} + E_{angle} + E_{dihedral}$$

$$E_{noncovalent} = E_{electrostatc} + E_{wander\ walls}$$

where the components of the covalent and noncolvalent contributions are given by the following summations:

The exact functional form of the potential function, or force field, depends on the particular simulation program being used. Generally the bond and angle terms are modeled as harmonic potentials centered around equilibrium bond-length values derived from experiment or theoretical calculations of electronic structure performed with software which does ab-initio type calculations such as Gaussian. For accurate reproduction of vibrational spectra, the Morse potential can be used instead, at computational cost. The dihedral or torsional terms typically have multiple minima and thus cannot be modeled as harmonic oscillators, though their specific functional form varies with the implementation. This class of terms may include "improper" dihedral terms, which function as correction factors for out-of-plane deviations (for example, they can be used to keep benzene rings planar).

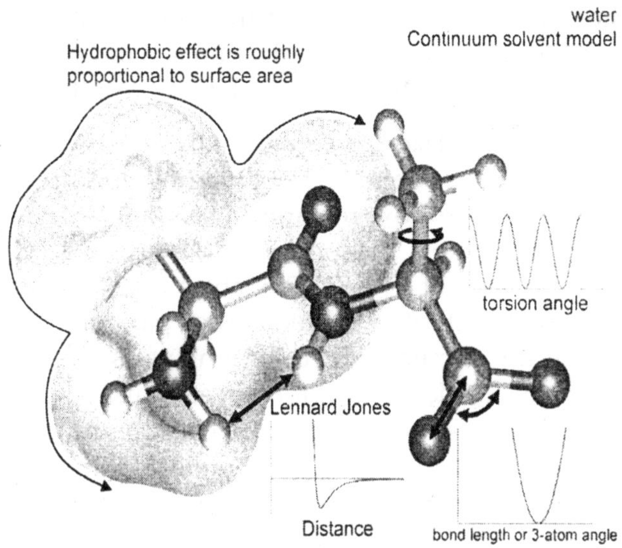

Figure 2. Molecular mechanics potential energy function with continuum solvent

The non-bonded terms are much more computationally costly to calculate in full, since a typical atom is bonded to only a few of its neighbors, but interacts with every other atom in the molecule. Fortunately the van der Waals term falls off rapidly - it is typically modeled using a "6-12 Lennard-Jones potential", which means that attractive forces fall off with distance as r^{-6} and repulsive forces as r^{-12}, where r represents the distance between two atoms. Generally a cutoff radius is used to speed up the calculation so that atom pairs whose distances are greater than the cutoff have a van der Waals interaction energy of zero.

The electrostatic terms are notoriously difficult to calculate well because they do not fall off rapidly with distance, and long-range electrostatic interactions are often important features of the system under study (especially for proteins). The basic functional form is the Coulomb potential, which only falls off as r^{-1}. A variety of methods are used to address this problem, the simplest being a cutoff radius similar to that used for the van der Waals terms. However, this introduces a sharp discontinuity between atoms inside and atoms outside the radius. Switching or scaling functions that modulate the apparent electrostatic energy are somewhat more accurate methods that multiply the calculated energy by a smoothly varying scaling factor from 0 to 1 at the outer and inner cutoff radii. Other more sophisticated but computationally intensive methods are known as particle mesh Ewald (PME) and the multipole algorithm.

In addition to the functional form of each energy term, a useful energy function must be assigned parameters for force constants, van der Waals multipliers, and other constant terms. These terms, together with the equilibrium bond, angle, and dihedral values, partial charge values, atomic masses and radii, and energy function definitions, are collectively known as a force field. Parameterization is typically done through agreement with experimental values and theoretical calculations results.

Each force field is parameterized to be internally consistent, but the parameters are generally not transferable from one force field to another. For example, the MM3 force field has 46 sets of parameters for oxygen and 18 sets for carbon (in different functional groups). MM3 has no parameters for lone pairs of electrons. On the other hand, the molecular recognition features of STR3DI32 enables it to use one set of parameters for each hybridization type of each atom (three sets of parameters for each of the second row elements, etc.), and it is parameterized for lone pairs of electrons. These molecular recognition features of STR3DI32 allow for an easy usage of data from MM3 files. However, starting with the simple Cartesian coordinate data, one would require considerable effort to produce a MM3 file. Most molecular modelers, and STR3DI32, can generate MM2, and other, files which can be read by MM3.

Molecular Mechanics

The prototypical Molecular Mechanics application is energy minimization. That is, the force field is used as an optimization criterion and the (local) minimum searched by an appropriate algorithm (e.g. steepest descent). Global energy optimization can be accomplished using simulated annealing, the Metropolis algorithm and other Monte Carlo methods, or using different deterministic methods of discrete or continuous optimization. The main aim of optimization methods is finding the lowest energy conformation of a molecule or identifying a set of low-energy conformers that are in equilibrium with each other. The force field represents only the enthalpic component of free energy, and only this component is included during energy minimization. However, the analysis of equilibrium between different states requires also conformational entropy be included, which is possible but rarely done.

Molecular mechanics potential energy functions have been used to calculate binding constants, protein folding kinetics, protonation equilibria, active site coordinates, and to design binding sites. Molecular Mechanics and Molecular Dynamics (MD) are related but different. Main purpose of MD is modeling of molecular motions, although it is also applied for optimization, for example using simulated annealing. MM implements more "static" energy minimization methods to study the potential energy surfaces of different molecular systems. However, MM can also provide important dynamic parameters, such as energy barriers between different conformers or steepness of a potential energy surface around a local minimum. MD and MM are usually based on the same classical force fields. But MD may also be based on quantum chemical methods like DFT. MM is also loosely used to define a set of techniques in molecular modeling.

Environment and Solvation

There are several ways of defining the environment surrounding the molecule or molecules of interest in molecular mechanics. A system can be simulated in vacuum (known as a gas-phase simulation) with no surrounding environment at all, but this is usually not desirable because it introduces artifacts in the molecular geometry, especially in charged molecules. Surface charges that would ordinarily interact with solvent molecules instead interact with each other, producing molecular conformations that are unlikely to be present in any other environment. The "best" way to solvate a system is to place explicit water molecules in the simulation box with the molecules of interest and treat the water molecules as interacting particles like those in the molecule.

A variety of water models exist with increasing levels of complexity, representing water as a simple hard sphere (a united-atom approach), as three separate particles with fixed bond angles, or even as four or five separate interaction centers to account for unpaired electrons on the oxygen atom. Unsurprisingly, the more complex the water model, the more computationally intensive the simulation. A compromise approach has been found in implicit solvation, which replaces the explicitly represented water molecules with a mathematical expression that reproduces the average behavior of water molecules (or other solvents such as lipids). This method is useful for preventing artifacts that arise from vacuum simulations and reproduces bulk solvent properties well, but cannot reproduce situations in which individual water molecules have interesting interactions with the molecules under study.

PROTEIN ENGINEERING

Protein engineering is the application of science, mathematics, and economics to the process of developing useful or valuable proteins. It is a young discipline, with much research currently taking place into the understanding of protein folding and protein recognition for protein design principles.There are two general strategies for protein engineering. The first is known as rational design, in which the scientist uses detailed knowledge of the structure and function of the protein to make desired changes. This has the advantage of being generally inexpensive and easy, since site-directed mutagenesis techniques are well-developed. However, there is a major drawback in that detailed structural knowledge of a protein is often unavailable, and even when it is available, it can be extremely difficult to predict the effects of various mutations. Computational protein design algorithms seek to identify amino acid sequences that have low energies for target structures. While the sequence-conformation space that needs to be searched is large, the most challenging requirement for computational protein design is a fast, yet accurate, energy function that can distinguish optimal sequences from similar suboptimal ones. Using computational methods, a protein with a novel fold has been designed, as well as sensors for un-natural molecules.

The second strategy is known as directed evolution. This is where random mutagenesis is applied to a protein, and a selection regime is used to pick out variants that have the desired qualities. Further rounds of mutation and selection are then applied. This method mimics natural evolution and generally produces superior results to rational design. An additional technique known as DNA shuffling mixes and matches pieces of successful variants in order to produce better results. This process mimics recombination that occurs naturally during sexual reproduction.

The great advantage of directed evolution techniques is that they require no prior structural knowledge of a protein, nor it is necessary to be able to predict what effect a given mutation will have. Indeed, the results of directed evolution experiments are often surprising in that desired changes are often caused by mutations that no one would have expected. The drawback is that they require high-throughput, which is not feasible for all proteins. Large amounts of recombinant DNA must be mutated and the products screened for desired qualities. The sheer number of variants often requires expensive robotic equipment to automate the process. Furthermore, not all desired activities can be easily screened for.

Rational design and directed evolution techniques are not mutually exclusive; good researchers will often apply both. In the future, more detailed knowledge of protein structure and function, as well as advancements in high-throughput technology, will greatly expand the capabilities of protein engineering. Eventually even unnatural amino acids may be incorporated thanks to a new method that allows the incorporation of novel amino acids in the genetic code.

Carbon Nanotubes

Carbon nanotubes (CNTs) are allotropes of carbon with a nanostructure that can have a length-to-diameter ratio greater than 1,000,000. These cylindrical carbon molecules have novel properties that make them potentially useful in many applications in nanotechnology, electronics, optics and other fields of materials science, as well as potential uses in architectural fields. They exhibit extraordinary strength and unique electrical properties, and are efficient conductors of heat. Inorganic nanotubes have also been synthesized.Nanotubes are members of the fullerene structural family, which also includes the spherical buckyballs. The cylindrical nanotube usually has at least one end capped with a hemisphere of the buckyball structure. Their name is derived from their size, since the diameter of a nanotube is in the order of a few nanometers (approximately

1/50,000th of the width of a human hair), while they can be up to several millimeters in length (as of 2008). Nanotubes are categorized as single-walled nanotubes (SWNTs) and multi-walled nanotubes (MWNTs).The nature of the bonding of a nanotube is described by applied quantum chemistry, specifically, orbital hybridization. The chemical bonding of nanotubes is composed entirely of sp2 bonds, similar to those of graphite. This bonding structure, which is stronger than the sp3 bonds found in diamonds, provides the molecules with their unique strength. Nanotubes naturally align themselves into "ropes" held together by Van der Waals forces. Under high pressure, nanotubes can merge together, trading some sp^2 bonds for sp^3 bonds, giving the possibility of producing strong, unlimited-length wires through high-pressure nanotube linking.

Single-walled Carbon Nanotubes

Most single-walled nanotubes (SWNT) have a diameter of close to 1 nanometer, with a tube length that can be many thousands of times longer. The structure of a SWNT can be conceptualized by wrapping a one-atom-thick layer of graphite called graphene into a seamless cylinder. The way the graphene sheet is wrapped is represented by a pair of indices (n,m) called the chiral vector. The integers n and m denote the number of unit vectors along two directions in the honeycomb crystal lattice of graphene. If m=0, the nanotubes are called "zigzag". If n=m, the nanotubes are called "armchair". Otherwise, they are called "chiral".

Single-walled nanotubes are a very important variety of carbon nanotube because they exhibit important electric properties that are not shared by the multi-walled carbon nanotube (MWNT) variants. Single-walled nanotubes are the most likely candidate for miniaturizing electronics beyond the micro electromechanical scale that is currently the basis of modern electronics. The most basic building block of these systems is the electric wire, and SWNTs can be excellent conductors.

One useful application of SWNTs is in the development of the first intramolecular field effect transistors (FETs). The production of the first intramolecular logic gate using SWNT FETs has recently become possible as well. To create a logic gate you must have both a p-FET and an n-FET. Because SWNTs are p-FETs when exposed to oxygen and n-FETs when unexposed to oxygen, it is possible to protect half of a SWNT from oxygen exposure, while exposing the other half to oxygen. This results in a single SWNT that acts as a NOT logic gate with both p and n-type FETs within the same molecule.

Single-walled nanotubes are still very expensive to produce, around $1500 per gram as of 2000, and the development of more affordable synthesis techniques is vital to the future of carbon nanotechnology. If cheaper means of synthesis cannot be discovered, it would make it financially impossible to apply this technology to commercial-scale applications. Several suppliers offer as-produced arc discharge SWNTs for ~$50-100 per gram as of 2007.

Multi-walled Carbon Nanotubes

Multi-walled nanotubes (MWNT) consist of multiple layers of graphite rolled in on themselves to form a tube shape. There are two models which can be used to describe the structures of multi-walled nanotubes. In the Russian Doll model, sheets of graphite are arranged in concentric cylinders, e.g. a (0,8) single-walled nanotube (SWNT) within a larger (0,10) single-walled nanotube. In the Parchment model, a single sheet of graphite is rolled in around itself, resembling a scroll of parchment or a rolled up newspaper. The interlayer distance in multi-walled nanotubes is close to the distance between graphene layers in graphite, approximately 3.3 Å (330 pm).

The special place of double-walled carbon nanotubes (DWNT) must be emphasized here because they combine very similar morphology and properties as compared to SWNT, while improving significantly their resistance to chemicals. This is especially important when functionalization is required (this means grafting of chemical functions at the surface of the nanotubes) to add new properties to the CNT.

In the case of SWNT, covalent functionalization will break some C=C double bonds, leaving "holes" in the structure on the nanotube and thus modifying both its mechanical and electrical properties. In the case of DWNT, only the outer wall is modified. DWNT synthesis on the gram-scale was first proposed in 2003 by the CCVD technique, from the selective reduction of oxides solid solutions in methane and hydrogen.

Fullerite

Fullerites are the solid-state manifestation of fullerenes and related compounds and materials. Being highly incompressible nanotube forms, polymerized single-walled nanotubes (P-SWNT) are a class of fullerites and are comparable to diamond in terms of hardness. However, due to the way that nanotubes intertwine, P-SWNTs don't have the corresponding crystal lattice that makes it possible to cut diamonds neatly. This same structure results in a less brittle material, as any impact that the structure sustains is spread out throughout the material.

Torus

A nanotorus is a theoretically described carbon nanotube bent into a torus (doughnut shape). Nanotori have many unique properties, such as magnetic moments 1000 times larger than previously expected for certain specific radii. Properties such as magnetic moment, thermal stability, etc. vary widely depending on radius of the torus and radius of the tube.

Nanobud

Carbon nanobuds are a newly discovered material combining two previously discovered allotropes of carbon: carbon nanotubes and fullerenes. In this new material fullerene-like "buds" are covalently bonded to the outer sidewalls of the underlying carbon nanotube. This hybrid material has useful properties of both fullerenes and carbon nanotubes. In particular, they have been found to be exceptionally good field emitters. In composite materials, the attached fullerene molecules may function as molecular anchors preventing slipping of the nanotubes, thus improving the composite's mechanical properties.

Properties

Kinetic

Multi-walled nanotubes, multiple concentric nanotubes precisely nested within one another, exhibit a striking telescoping property whereby an inner nanotube core may slide, almost without friction, within its outer nanotube shell thus creating an atomically perfect linear or rotational bearing. This is one of the first true examples of molecular nanotechnology, the precise positioning of atoms to create useful machines. Already this property has been utilized to create the world's smallest rotational motor. Future applications such as a gigahertz mechanical oscillator are also envisaged.

Electrical

Because of the symmetry and unique electronic structure of graphene, the structure of a nanotube strongly affects its electrical properties. For a given (n,m) nanotube, if n - m is a multiple of 3, then the nanotube is metallic, otherwise the nanotube is a semiconductor. Thus all armchair (n=m) nanotubes are metallic, and nanotubes (5,0), (6,4), (9,1), etc. are semiconducting. In theory, metallic nanotubes can have an electrical current density more than 1,000 times greater than metals such as silver and copper.

Thermal

All nanotubes are expected to be very good thermal conductors along the tube, exhibiting a property known as "ballistic conduction," but good insulators laterally to the tube axis. It is predicted that carbon nanotubes will be able to transmit up to 6000 watts per meter per Kelvin at room temperature; compare this to copper, a metal well-known for its good thermal conductivity, which only transmits 385 $W \bullet m^{-1} \bullet K^{-1}$. The temperature stability of carbon nanotubes is estimated to be up to 2800 degrees Celsius in vacuum and about 750 degrees Celsius in air.

Defects

As with any material, the existence of defects affects the material properties. Defects can occur in the form of atomic vacancies. High levels of such defects can lower the tensile strength by up to 85%. Another form of defect that may occur in carbon nanotubes is known as the Stone Wales defect, which creates a pentagon and heptagon pair by rearrangement of the bonds. Because of the very small structure of CNTs, the tensile strength of the tube is dependent on the weakest segment of it in a similar manner to a chain, where a defect in a single link diminishes the strength of the entire chain.

The tube's electrical properties are also affected by the presence of defects. A common result is the lowered conductivity through the defective region of the tube. Some defect formation in armchair-type tubes (which can conduct electricity) can cause the region surrounding that defect to become semiconducting. Furthermore single monoatomic vacancies induce magnetic properties.The tube's thermal properties are heavily affected by defects. Such defects lead to phonon scattering, which in turn increases the relaxation rate of the phonons. This reduces the mean free path, and reduces the thermal conductivity of nanotube structures. Phonon transport simulations indicate that substitutional defects such as nitrogen or boron will primarily lead to scattering of high frequency optical phonons. However, larger scale defects such as Stone Wales defects cause phonon scattering over a wide range of frequencies, leading to a greater reduction in thermal conductivity.

One-Dimensional Transport

Due to their nanoscale dimensions, electron transport in carbon nanotubes will take place through quantum effects and will only propagate along the axis of the tube. Because of this special transport property, carbon nanotubes are frequently referred to as "one-dimensional" in scientific articles.

Toxicity

Determining the toxicity of carbon nanotubes has been one of the most pressing questions in Nanotechnology. Unfortunately such research has only just begun and the data are still fragmentary and subject to criticisms. Preliminary results highlight the difficulties in evaluating the toxicity of this heterogeneous material. Parameters such as structure, size distribution, surface area, surface chemistry, surface charge, and agglomeration state as well as purity of the samples, have considerable impact on the reactivity of carbon nanotubes. However, available data clearly show that, under some conditions, nanotubes can cross the membrane barriers and suggests that if raw materials reach the organs they can induce harmful effects as inflammatory and fibrotic reactions.The needle-like fiber shape of CNTs, similar to asbestos fibers, raises fears that widespread use of carbon nanotubes may lead to mesothelioma, cancer of the lining of the lungs caused by exposure to asbestos. A recently published pilot study supports this prediction. Scientists exposed the mesothelial lining of the body cavity of mice, as a surrogate for the mesothelial lining of the chest cavity, to long multiwalled carbon nanotubes and observed asbestos-like, length-dependent, pathogenic behavior which included inflammation and formation of lesions known as granulomas.

Synthesis

Techniques have been developed to produce nanotubes in sizeable quantities, including arc discharge, laser ablation, high pressure carbon monoxide (HiPCO), and chemical vapor deposition (CVD). Most of these processes take place in vacuum or with process gases. CVD growth of CNTs can take place in vacuum or at atmospheric pressure. Large quantities of nanotubes can be synthesized by these methods; advances in catalysis and continuous growth processes are making CNTs more commercially viable.

Arc discharge

Nanotubes were observed in 1991 in the carbon soot of graphite electrodes during an arc discharge, by using a current of 100 amps, that was intended to produce fullerenes. However the first macroscopic production of carbon nanotubes was made in 1992 by two researchers at NEC's Fundamental Research Laboratory. The method used was the same as in 1991. During this process, the carbon contained in the negative electrode sublimates because of the high temperatures caused by the discharge. Because nanotubes were initially discovered using this technique, it has been the most widely used method of nanotube synthesis.

Laser ablation

In the laser ablation process, a pulsed laser vaporizes a graphite target in a high temperature reactor while an inert gas is bled into the chamber. The nanotubes develop on the cooler surfaces of the reactor, as the vaporized carbon condenses. A water-cooled surface may be included in the system to collect the nanotubes.It was invented by Richard Smalley and co-workers at Rice University, who at the time of the discovery of carbon nanotubes, were blasting metals with the laser to produce various metal molecules. When they heard of the discovery they substituted the metals with graphite to create multi-walled carbon nanotubes. This method has a yield of around 70% and produces primarily single-walled carbon nanotubes with a controllable diameter determined by the reaction temperature. However, it is more expensive than either arc discharge or chemical vapor deposition.

Chemical vapor deposition (CVD)

The catalytic vapor phase deposition of carbon was first reported in 1959, but it was not until 1993 that carbon nanotubes could be formed by this process. In 2007, researchers at the University of Cincinnati (UC) developed a process to grow 18 mm long aligned carbon nanotube arrays on a FirstNano ET3000 carbon nanotube growth system.During CVD, a substrate is prepared with a layer of metal catalyst particles, most commonly nickel, cobalt, iron, or a combination. The metal nanoparticles can also be produced by other ways, including reduction of oxides or oxides solid solutions. The diameters of the nanotubes that are to be grown are related to the size of the metal particles. This can be controlled by patterned (or masked) deposition of the metal, annealing, or by plasma etching of a metal layer. The substrate is heated to approximately 700°C. To initiate the growth of nanotubes, two gases are bled into the reactor: a process gas (such as ammonia, nitrogen, hydrogen, etc.) and a carbon-containing gas (such as acetylene, ethylene, ethanol, methane, etc.). Nanotubes grow at the sites of the metal catalyst; the carbon-containing gas is broken apart at the surface of the catalyst particle, and the carbon is transported to the edges of the particle, where it forms the nanotubes. This mechanism is still under discussion. The catalyst particles can stay at the tips of the growing nanotube during the growth process, or remain at the nanotube base, depending on the adhesion between the catalyst particle and the substrate.

CVD is a common method for the commercial production of carbon nanotubes. For this purpose, the metal nanoparticles will be carefully mixed with a catalyst support (e.g., MgO, Al_2O_3, etc) to increase the specific surface area for higher yield of the catalytic reaction of the carbon feedstock with the metal particles. One issue in this synthesis route is the removal of the catalyst support via an acid treatment, which sometimes could destroy the original structure of the carbon nanotubes. However, alternative catalyst supports that are soluble in water have been shown to be effective for nanotube growth.

If a plasma is generated by the application of a strong electric field during the growth process (plasma enhanced chemical vapor deposition), then the nanotube growth will follow the direction of the electric field. By properly adjusting the geometry of the reactor it is possible to synthesize vertically aligned carbon nanotubes (i.e., perpendicular to the substrate), a morphology that has been of interest to researchers interested in the electron emission from nanotubes. Without the plasma, the resulting nanotubes are often randomly oriented. Under certain reaction conditions, even in the absence of a plasma, closely spaced nanotubes will maintain a vertical growth direction resulting in a dense array of tubes resembling a carpet or forest. Of the various means for nanotube synthesis, CVD shows the most promise for industrial scale deposition in terms of its price/unit ratio. There are additional advantages to the CVD synthesis of nanotubes. Unlike the above methods, CVD is capable of growing nanotubes directly on a desired substrate, whereas the nanotubes must be collected in the other growth techniques. The growth sites are controllable by careful deposition of the catalyst. Additionally, no other growth methods have been developed to produce vertically aligned nanotubes.

Natural, incidental, and controlled flame environments

Fullerenes and carbon nanotubes are not necessarily products of high-tech laboratories; they are commonly formed in such mundane places as ordinary flames, produced by burning methane, ethylene, and benzene, and they have been found in soot from both indoor and outdoor air. However, these naturally occurring varieties can be highly irregular in size and quality because the environment in which they are produced is often highly uncontrolled. Thus, although they can be used in some applications, they can lack in the high degree of uniformity necessary to meet many needs of both research and industry. Recent efforts have focused on producing more uniform carbon nanotubes in controlled flame environments. Nano-C, Inc of Westwood, Massachusetts, is producing flame synthesized single-walled carbon nanotubes. This method has promise for large scale, low cost nanotube synthesis, though it must compete with rapidly developing large scale CVD production.

Potential and Current Applications

The strength and flexibility of carbon nanotubes makes them of potential use in controlling other nanoscale structures, which suggests they will have an important role in nanotechnology engineering. The highest tensile strength an individual multi-walled carbon nanotube has been tested to be is 63 GPa.

Structural

Because of the great mechanical properties of the carbon nanotubule, a variety of structures have been proposed ranging from everyday items like clothes and sports gear to combat jackets and space elevators. However, the space elevator will require further efforts in refining carbon nanotube technology, as the practical tensile strength of carbon nanotubes can still be greatly improved.For perspective, outstanding breakthroughs have already been made. Pioneering work lead by Ray H. Baughman at the NanoTech Institute has shown that single and multi-walled nanotubes can produce materials with toughness un-matched in the man-made and natural worlds.Recent research by James D. Iverson and Brad C. Edwards has revealed the possibility of cross-linking CNT molecules prior to incorporation in a polymer matrix to form a super high strength composite material. This CNT composite could have a tensile strength on the order of 20 million psi (138 GPa, for 106 MN•m•kg^{-1}), potentially revolutionizing many aspects of engineering design where low weight and high strength is required.

In electrical circuits

Carbon nanotubes have many properties-from their unique dimensions to an unusual current conduction mechanism-that make them ideal components of electrical circuits. For example, they have shown to exhibit strong electron-phonon resonances, which indicate that under certain direct current (dc) bias and doping conditions their current and the average electron velocity, as well as the electron concentration on the tube oscillate at terahertz frequencies. These resonances can be used to make terahertz sources or sensors. One major obstacle to realization of nanotubes has been the lack of technology for mass production. However, in 2001 IBM researchers demonstrated how nanotube transistors can be grown in bulk, not very differently from silicon transistors. The process they used is called "constructive destruction" which includes the automatic destruction of defective nanotubes on the wafer.This has since then been developed further and single-chip wafers with over ten billion correctly aligned nanotube junctions have been created. In addition it has been demonstrated that incorrectly aligned nanotubes can be removed automatically using standard photolithography equipment.The first nanotube integrated memory circuit was made in 2004. One of the main challenges has been regulating the conductivity of nanotubes. Depending on subtle surface features a nanotube may act as a plain conductor or as a semiconductor. A fully automated method has however been developed to remove non-semiconductor tubes.

Other applications

Nanomix Inc was the first to put on the market an electronic device that integrated carbon nanotubes on a silicon platform, in May 2005. It was a Hydrogen sensor. Since then Nanomix has been patenting many such sensor applications such as in the field of carbon dioxide, Nitrous Oxide, glucose, DNA detection etc...Eikos Inc of Franklin, Massachusetts and Unidym Inc. of Silicon Valley, California are developing transparent, electrically conductive films of carbon nanotubes to replace indium tin oxide (ITO). Carbon nanotube films are substantially more mechanically robust than ITO films, making them ideal for high reliability touch screens and flexible displays. Printable water-based inks of carbon nanotubes are desired to enable the production of these films to replace ITO. Nanotube films show promise for use in displays for computers, cell phones, PDAs, and ATMs.

Carbon nanotubes are said to have the strength of diamond, and research is being made into weaving them into clothes to create stab-proof and bulletproof clothing. The nanotubes would effectively stop the bullet from penetrating the body but the force and velocity of the bullet would be likely to cause broken bones and internal bleeding.A flywheel made of carbon nanotubes could be spun at extremely high velocity on a floating magnetic axis, and potentially store energy at a density approaching that of conventional fossil fuels. Since energy can be added to and removed from flywheels very efficiently in the form of electricity, this might offer a way of storing electricity, making the electrical grid more efficient and variable power suppliers (like wind turbines) more useful in meeting energy needs. The practicality of this depends heavily upon the cost of making massive, unbroken nanotube structures, and their failure rate under stress.

MOLECULAR MODELLING

A molecular model is a physical model that represents molecules and their processes. The creation of mathematical models of molecular properties and behaviour is molecular modelling, and their graphical depiction is molecular graphics, but these topics are closely linked and each uses techniques from the others. The electronic structure is often also omitted or represented in a highly simplified way.Physical models of atomistic systems have played an important role in understanding chemistry and generating and testing hypotheses. Most commonly there is an explicit representation of atoms, though other approaches such as soap films and other continuous media have been useful. There are several motivations for creating physical models:

- as pedagogic tools for students or those unfamiliar with atomistic structures;
- as objects to generate or test theories (e.g., the structure of DNA);
- as analogue computers (e.g., for measuring distances and angles in flexible systems);
- as aesthetically pleasing objects on the boundary of art and science.

The construction of physical models is often a creative act, and many bespoke examples have been carefully created in the workshops of science departments. There is a very wide range of approaches to physical modelling, and this article lists only the most common or historically important. The main strategies are:

- bespoke construction of a single model;
- use of common materials (plasticine, matchsticks) or children's toys (Tinkertoy(TM), Meccano, Lego, etc.);
- re-use of generic components in kits.

Models encompass a wide range of degrees of precision and engineering: some models such as J.D. Bernal's water are conceptual, while the macromodels of Pauling and Crick and Watson were created with much greater precision.Molecular models have inspired molecular graphics, initially in textbooks and research articles and more recently on computers. Molecular graphics has replaced some functions of physical molecular models, but physical kits continue to be very popular and are sold in large numbers. Their unique strengths include:

- cheapness and portability;
- immediate tactile and visual messages;
- easy interactivity for many processes (e.g., conformational analysis and pseudorotation).

In the 1600s, Johannes Kepler speculated on the symmetry of snowflakes and also on the close packing of spherical objects such as fruit (this problem remained unsolved until very recently). The symmetrical arrangement of closely packed spheres informed theories of molecular structure in the late 1800s, and many theories of crystallography and solid state inorganic structure used collections of equal and unequal spheres to simulate packing and predict structure.

John Dalton represented compounds as aggregations of circular atoms, and although Loschmidt did not create physical models, his diagrams based on circles are two-dimensional analogues of later models. Hofmann is credited with the first physical molecular model around 1860 (Fig. 3).

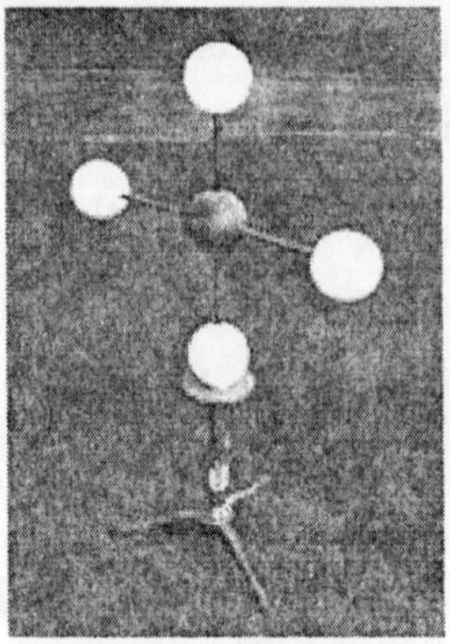

Figure 3. Hofmann's model for methane

Note how the size of the carbon appears smaller than the hydrogen. The importance of stereochemistry was not then recognised and the model is essentially topological (it should be a 3-dimensional tetrahedron).

Models-based on Spheres

Robert Hooke proposed a relationship between crystals and the packing of spheres. R. Haüy argued that the structures of crystals involved regular lattices of repeating units with shapes similar to the macroscopic crystal. Barlow, who jointly developed the theories of space groups, proposed models of crystals based on sphere packings. The binary compounds sodium chloride (NaCl) and caesium chloride (CsCl) have cubic structures but have different space groups. This can be rationalised in terms of close packing of spheres of different sizes. For example, NaCl can be described as close-packed chloride ions (in a face-centered cubic lattice) with sodium ions in the octahedral holes. After the development of X-ray crystallography as a tool for determining crystal structures, many laboratories built models based on spheres. With the development of plastic or polystyrene balls it is now easy to create such models.

Models-based on Ball-and-stick

The concept of the chemical bond as a direct link between atoms can be modelled by linking balls (atoms) with sticks/rods (bonds). This has been extremely popular and is still widely used today. Initially atoms were made of spherical wooden balls with specially drilled holes for rods. Thus carbon can be represented as a sphere with four holes at the tetrahedral angles $\cos^{-1}(-1/3) \approx 109.47°$.A problem with rigid bonds and holes is that systems with arbitrary angles could not be built. This can be overcome with flexible bonds, originally helical springs but now usually plastic. This also allows double and triple bonds to be approximated by multiple single bonds (Fig. 4). Figure 2 represents a ball-and-stick model of proline. The balls have colours: black represents carbon (C); red, oxygen (O); blue, nitrogen (N); and white, hydrogen (H). Each ball is drilled with as many holes as its conventional valence (C: 4; N: 3; O: 2; H: 1) directed towards the vertices of a tetrahedron. Single bonds are represented by (fairly) rigid grey rods. Double and triple bonds use two longer flexible bonds which restrict rotation and support conventional cis/trans stereochemistry.

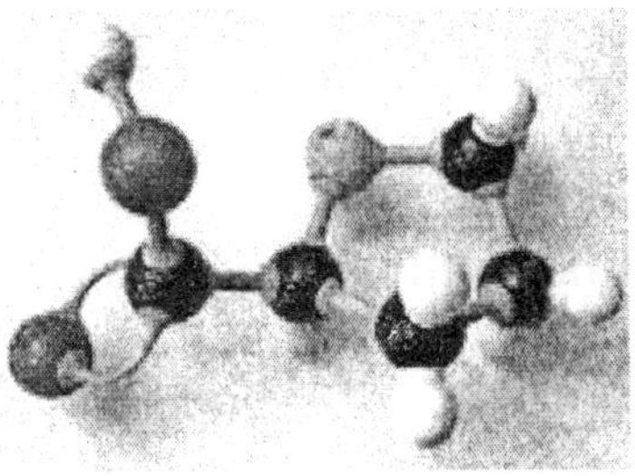

Figure 4. A modern plastic ball and stick model.

However, most molecules require holes at other angles and specialist companies manufacture kits and bespoke models. One of the earlier companies was Woosters at Bottisham, Cambridgeshire, UK. Besides tetrahedral, trigonal and octahedral holes, there were all-purpose balls with 24 holes. These models allowed rotation about the single rod bonds, which could be both an advantage (showing molecular flexibility) and a disadvantage (models are floppy). The approximate scale was 5 cm per ångström (0.5 m/nm or 500,000,000:1), but was not consistent over all elements.Arnold Beevers in Edinburgh (now operating as Miramodus) created small models using PMMA balls and stainless steel rods. By using individually drilled balls with precise bond angles and bond lengths in these models, large crystal structures to be accurately created, but with light and rigid form. Figure 5 shows a unit cell of ruby in this style.

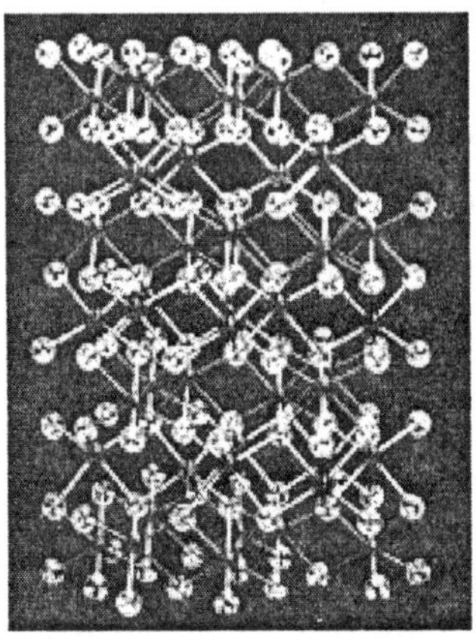

Figure 5. Beevers model of ruby (Cr-doped corundum).

Skeletal Models

Crick and Watson's DNA model and the protein-building kits of Kendrew were among the first skeletal models. These were based on atomic components where the valences were represented by rods; the atoms were points at the intersections. Bonds were created by linking components with tubular connectors with locking screws.Andre Dreiding introduced a molecular modelling kit which dispensed with the connectors. A given atom would have solid and hollow valence spikes. The solid rods clicked into the tubes forming a bond, usually with free rotation. These were and are very widely used in organic chemistry departments and were made so accurately that interatomic measurements could be made by ruler.More recently, inexpensive plastic models (such as Orbit) use a similar principle. A small plastic sphere has protuberances onto which plastic tubes can be fitted. The flexibility of the plastic means that distorted geometries can be made.

Polyhedral Models

Many inorganic solids consist of atoms surrounded by a coordination sphere of electronegative atoms (e.g. PO_4 tetrahedra, TiO_6 octahedra). Structures can be modelled by gluing together polyhedra made of paper or plastic.

Composite Models

A good example of composite models is the Nicholson approach, widely used from the late 1970s for building models of biological macromolecules. The components are primarily amino acids and nucleic acids with preformed residues representing groups of atoms. Many of these atoms are directly moulded into the template, and fit together by pushing plastic stubs into small holes. The plastic grips well and makes bonds difficult to rotate, so that arbitrary torsion angles can be set and retain their value. The conformations of the backbone and side chains are determined by pre-computing the torsion angles and then adjusting the model with a protractor.

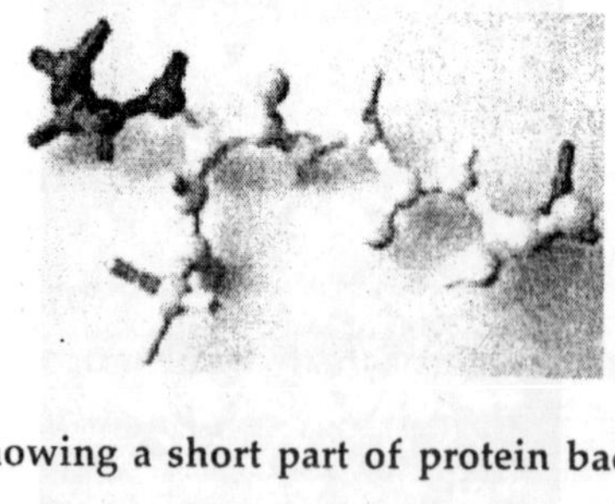

Figure 4. A Nicholson model, showing a short part of protein backbone (white) with side chains.

The plastic is white and can be painted to distinguish between O and N atoms. Hydrogen atoms are normally implicit and modelled by snipping off the spokes. A model of a typical protein with approximately 300 residues could take a month to build. It was common for laboratories to build a model for each protein solved. By 2005, so many protein structures were being determined that relatively few models were made.

MOLECULAR MODELLING

Molecular modelling is a collective term that refers to theoretical methods and computational techniques to model or mimic the behaviour of molecules. The techniques are used in the fields of computational chemistry, computational biology and materials science for studying molecular systems ranging from small chemical systems to large biological molecules and material assemblies. The simplest calculations can be performed by hand, but inevitably computers are required to perform molecular modelling of any reasonably sized system.

The common feature of molecular modelling techniques is the atomistic level description of the molecular systems; the lowest level of information is individual atoms (or a small group of atoms). This is in contrast to quantum chemistry (also known as electronic structure calculations) where electrons are considered explicitly. The benefit of molecular modelling is that it reduces the complexity of the system, allowing many more particles (atoms) to be considered during simulations.

Molecular mechanics is one aspect of molecular modelling, as it is refers to the use of classical mechanics/Newtonian mechanics to describe the physical basis behind the models. Molecular models typically describe atoms (nucleus and electrons collectively) as point charges with an associated mass. The interactions between neighbouring atoms are described by spring-like interactions (representing chemical bonds) and van der Waals forces.

The Lennard-Jones potential is commonly used to describe van der Waals forces. The electrostatic interactions are computed based on Coulomb's law. Atoms are assigned coordinates in Cartesian space or in internal coordinates, and can also be assigned velocities in dynamical simulations. The atomic velocities are related to the temperature of the system, a macroscopic quantity. The collective mathematical expression is known as a potential function and is related to the system internal energy (U), a thermodynamic quantity equal to the sum of potential and kinetic energies. Methods which minimize the potential energy are known as energy minimization techniques (e.g., steepest descent and conjugate gradient), while methods that model the behaviour of the system with propagation of time are known as molecular dynamics.

$$E = E_{bonds} + E_{angle} + E_{dihedral} + E_{non-bonded}$$

$$E_{non-bonded} = E_{electrostatic} + E_{vanderWaals}$$

This function, referred to as a potential function, computes the molecular potential energy as a sum of energy terms that describe the deviation of bond lengths, bond angles and torsion angles away from equilibrium values, plus terms for non-bonded pairs of atoms describing van der Waals and electrostatic interactions. The set of parameters consisting of equilibrium bond lengths, bond angles, partial charge values, force constants and van der Waals parameters are collectively known as a force field. Different implementations of molecular mechanics use slightly different mathematical expressions, and therefore, different constants for the potential function. The common force fields in use today have been developed by using high level quantum calculations and/or fitting to experimental data. The technique known as energy minimization is used to find positions of zero gradient for all atoms, in other words, a local energy minimum. Lower energy states are more stable and are commonly investigated because of their role in chemical and biological processes.

A molecular dynamics simulation, on the other hand, computes the behaviour of a system as a function of time. It involves solving Newton's laws of motion, principally the second law, F = ma. Integration of Newton's laws of motion, using different integration algorithms, leads to atomic trajectories in space and time. The force on an atom is defined as the negative gradient of the potential energy function. The energy minimization technique is useful for obtaining a static picture for comparing between states of similar systems, while molecular dynamics provides information about the dynamic processes with the intrinsic inclusion of temperature effects.

Molecules can be modelled either in vacuum or in the presence of a solvent such as water. Simulations of systems in vacuum are referred to as gas-phase simulations, while those that include the presence of solvent molecules are referred to as explicit solvent simulations. In another type of simulation, the effect of solvent is estimated using an empirical mathematical expression; these are known as implicit solvation simulations.Molecular modelling methods are now routinely used to investigate the structure, dynamics and thermodynamics of inorganic, biological, and polymeric systems. The types of biological activity that have been investigated using molecular modelling include protein folding, enzyme catalysis, protein stability, conformational changes associated with biomolecular function, and molecular recognition of proteins, DNA, and membrane complexes.

MOLECULAR GRAPHICS

Molecular graphics (MG) is the discipline and philosophy of studying molecules and their properties through graphical representation. IUPAC limits the definition to representations on a "graphical display device". Ever since Dalton's atoms and Kekule's benzene, there has been a rich history of hand-drawn atoms and molecules, and these representations have had an important influence on modern molecular graphics. This article concentrates on the use of computers to create molecular graphics. There has been a long tradition of creating molecular models from physical materials. Perhaps the best known is Crick and Watson's model of DNA built from rods and planar sheets, but the most widely used approach is to represent all atoms and bonds explicitly using the "ball and stick" approach. This can demonstrate a wide range of properties, such as shape, relative size, and flexibility. Many chemistry courses expect that students will have access to ball and stick models. One goal of mainstream molecular graphics has been to represent the "ball and stick" model as realistically as possible and to couple this with calculations of molecular properties.

Comparison of Physical Models

Physical models and computer models have partially complementary strengths and weaknesses. Physical models can be used by those without access to a computer and now can be made cheaply out of plastic materials. Their tactile and visual aspects cannot be easily reproduced by computers (although haptic devices have occasionally been built). On a computer screen, the flexibility of molecules is also difficult to appreciate; illustrating the pseudorotation of cyclohexane is a good example of the value of mechanical models.However, it is difficult to build large physical molecules, and all-atom physical models of even simple proteins could take weeks or months to build. Moreover, physical models are not robust and they decay over time. Molecular graphics is particularly valuable for representing global and local properties of molecules, such as electrostatic potential.

Initially the rendering was on early CRT screens or through plotters drawing on paper. Molecular structures have always been an attractive choice for developing new computer graphics tools, since the input data are easy to create and the results are usually highly appealing. The first example of MG was a display of a protein molecule by Cyrus Levinthal and Robert Langridge. Among the milestones in high-performance MG was the work of Nelson Max in "realistic" rendering of macromolecules using reflecting spheres.By about 1980 many laboratories both in academia and industry had recognized the power of the computer to analyse and predict the properties of molecules, especially in materials science and the pharmaceutical industry. The discipline was often called "molecular graphics" and in 1982 a group of academics and industrialists in the UK set up the Molecular Graphics Society (MGS). Initially much of the technology concentrated either on high-performance 3D graphics, including interactive rotation or 3D rendering of atoms as spheres (sometimes with radiosity). During the 1980s a number of programs for calculating molecular properties (such as molecular dynamics and quantum mechanics) became available and the term "molecular graphics" often included these. As a result the MGS has now changed its name to the Molecular Graphics and Modelling Society (MGMS).The requirements of macromolecular crystallography also drove MG because the traditional techniques of physical model-building could not scale. Alwyn Jones' FRODO program (and later "O") were developed to overlay the molecular electron density determined from X-ray crystallography and the hypothetical molecular structure.

Science and Technology in Molecular Graphics

Both computer technology and graphic arts have contributed to molecular graphics. The development of structural biology in the 1950s led to a requirement to display molecules with thousands of atoms. The existing computer technology was limited in power, and in any case a naive depiction of all atoms left viewers overwhelmed. Most systems therefore used conventions where information was implicit or stylistic. Two vectors meeting at a point implied an atom or (in macromolecules) a complete residue (10-20 atoms).The macromolecular approach was popularized by Dickerson and Geis' presentation of proteins and the graphic work of Jane Richardson through high-quality hand-drawn diagrams. In this they strove to capture the intrinsic 'meaning' of the molecule. This search for the "messages in the molecule" has always accompanied the increasing power of computer graphics processing. Typically the depiction would concentrate on specific areas of the molecule (such as the active site) and this might have different colours or more detail in the number of explicit atoms or the type of depiction (e.g., spheres for atoms).

In some cases the limitations of technology have led to serendipitous methods for rendering. Most early graphics devices used vector graphics, which meant that rendering spheres and surfaces was impossible. Michael Connolly's program "MS" calculated points on the surface-accessible surface of a molecule, and the points were rendered as dots with good visibility using the new vector graphics technology, such as the Evans and Sutherland PS300 series. Thin sections ("slabs") through the structural display showed very clearly the complementarity of the surfaces for molecules binding to active sites, and the "Connolly surface" became a universal metaphor.The relationship between the art and science of molecular graphics is shown in the exhibitions sponsored by the Molecular Graphics Society. Some exhibits are created with molecular graphics programs alone, while others are collages, or involve physical materials. An example from Mike Hann, inspired by Magritte's painting Ceci n'est pas une pipe, uses an image of a salmeterol molecule.

For many years the size of atoms has been approximated by mechanical models (CPK), where the atoms have been represented by plastic spheres whose radius (van der Waals radius) describes a sphere within which "most" of the electron density can be found. These spheres could be clicked together to show the steric aspects of the molecule rather than the positions of the nuclei. Since the atomic radii are only slightly less than the distance between bonded atoms, the iconic spheres intersect, and in the CPK models, this was achieved by planar truncations along the bonding directions, the section being circular. When raster graphics became affordable, one of the common approaches was to replicate CPK models in silico. It is relatively straightforward to calculate the circles of intersection, but more complex to represent a model with hidden surface removal. A useful side product is that a conventional value for the molecular volume can be calculated.The use of spheres is often for convenience, being limited both by graphics libraries and the additional effort required to compute complete electronic density or other space-filling quantities. It is now relatively common to see images of isosurfaces that have been coloured to show quantities such as electrostatic potential. The commonest isosurfaces are the Connolly surface, or the volume within which a given proportion of the electron density lies.

Molecular graphics has always pushed the limits of display technology, and has seen a number of cycles of integration and separation of compute-host and display. Early systems like Project MAC were bespoke and unique, but in the 1970s the MMS-X and similar systems used (relatively) low-cost terminals, such as the Tektronix 4014 series, often over dial-up lines to multi-user hosts. The devices could only display static pictures but, were able to evangelize MG. In the late 1970s, it was possible for departments (such as crystallography) to afford their own hosts (e.g., PDP-11) and to attach a display (such as Evans & Sutherland's MPS) directly to the bus. The display list was kept on the host, and interactivity was good since updates were rapidly reflected in the display-at the cost of reducing most machines to a single-user system.

In the early 1980s, Evans & Sutherland (E&S) decoupled their PS300 display, which contained its own display information transformable through a dataflow architecture. Complex graphical objects could be downloaded over a serial line (e.g. 9600 baud) and then manipulated without impact on the host. The architecture was excellent for high performance display but very inconvenient for domain-specific calculations, such as electron-density fitting and energy calculations. Many crystallographers and modellers spent arduous months trying to fit such activities into this architecture.

The benefits for MG were considerable, but by the later 1980s, UNIX workstations such as Sun-3 with raster graphics (initially at a resolution of 256 by 256) had started to appear. Computer-assisted drug design in particular required raster graphics for the display of computed properties such as atomic charge and electrostatic potential. Although E&S had a high-end range of raster graphics (primarily aimed at the aerospace industry) they failed to respond to the low-end market challenge where single users, rather than engineering departments, bought workstations. As a result the market for MG displays passed to Silicon Graphics, coupled with the development of minisupercomputers (e.g., CONVEX and Alliant) which were affordable for well-supported MG laboratories. Silicon Graphics provided a graphics language, IrisGL, which was easier to use and more productive than the PS300 architecture. Commercial companies (e.g., Biosym, Polygen/MSI) ported their code to Silicon Graphics, and by the early 1990s, this was the "industry standard".

Stereoscopic displays were developed based on liquid crystal polarized spectacles, and while this had been very expensive on the PS300, it now became a commodity item. A common alternative was to add a polarizable screen to the front of the display and to provide viewers with extremely cheap spectacles with orthogonal polarization for separate eyes. With projectors such as Barco, it was possible to project stereoscopic display onto special silvered screens and supply an audience of hundreds with spectacles. In this way molecular graphics became universally known within large sectors of chemical and biochemical science, especially in the pharmaceutical industry. Because the backgrounds of many displays were black by default, it was common for modelling sessions and lectures to be held with almost all lighting turned off.

In the last decade almost all of this technology has become commoditized. IrisGL evolved to OpenGL so that molecular graphics can be run on any machine. In 1992, Roger Sayle released his RasMol program into the public domain. RasMol contained a very high-performance molecular renderer that ran on Unix/X Window, and Sayle later ported this to the Windows and Macintosh platforms. The Richardsons developed Kinemage, which was also multi-platform. By specifying the chemical MIME type, molecular models could be served over the Internet, so that for the first time MG could be distributed at zero cost regardless of platform. In 1995, Birkbeck College's crystallography department used this to run "Principles of Protein Structure", the first multimedia course on the Internet, which reached 100 to 200 scientists.

MG continues to see innovation that balances technology and art, and currently zero-cost or open source programs such as PyMOL and Jmol have very wide use and acceptance. Recently the wide spread diffusion of advanced graphics hardware, has improved the rendering capabilities of the visualization tools. The capabilities of current shading languages allow the inclusion of advanced graphic effects (like ambient occlusion, cast shadows and non-photorealistic rendering techniques) in the interactive visualization of molecules. These graphic effects, beside being eye candy, can improve the comprehension of the three dimensional shapes of the molecules. An example of the effects that can be achieved exploiting recent graphics hardware can be seen in the simple open source visualization system QuteMol.

Drawing molecules requires a transformation between molecular coordinates (usually, but not always, in Angstrom units) and the screen. Because many molecules are chiral it is essential that the handedness of the system (almost always right-handed) is preserved. In molecular graphics the origin (0, 0) is usually at the lower left, while in many computer systems the origin is at top left. If the z-coordinate is out of the screen (towards the viewer) the molecule will be referred to right-handed axes, while the screen display will be left-handed.

Molecular transformations normally require:

- scaling of the display (but not the molecule).
- translations of the molecule and objects on the screen.
- rotations about points and lines.

Monte Carlo molecular Modeling

Monte Carlo molecular modeling is the application of Monte Carlo methods to molecular problems. These problems can also be modeled by the molecular dynamics method. The difference is that this approach relies on statistical mechanics rather than molecular dynamics. Instead of trying to reproduce the dynamics of a system, it generates states according to appropriate Boltzmann probabilities. Thus, it is the application of the Metropolis Monte Carlo simulation to molecular systems. It is therefore also a particular subset of the more general Monte Carlo method in statistical physics. It employs a Markov chain procedure in order to determine a new state for a system from a previous one. According to its stochastic nature, this new state is accepted at random. Each trial usually counts as a move. The avoidance of dynamics restricts the method to studies of static quantities only, but the freedom to choose moves makes the method very flexible. These moves must only satisfy a basic condition of balance in order equilibrium be properly described, but detailed balance, a stronger condition, is usually imposed when designing new algorithms. An additional advantage is that some systems, such as the Ising model, lack a dynamical description and are only defined by an energy prescription; for these the Monte Carlo approach is the only one feasible. The great success of this method in statistical mechanics has led to various generalizations such as the method of simulated annealing for optimization, in which a fictitious temperature is introduced and then gradually lowered.

Space-filling Model

Space-filling models (also known as calotte models or CPK models by Corey, Pauling and Koltun), like ball-and-stick models, belong to the 3D or spatial molecular models, which serve to display the structure of chemical products and substances or biomolecules.The calotte model is a further development of the ball-and-stick model, with which the configuration of molecules can be vividly presented. With it, the atoms of individual elements are represented by multicolored spherical sections. Atom size, bond angle and bond length correspond to the actual relationships. The surface usually represents the van der Waals radius of the individual atoms.

MOLECULAR DOCKING

In the field of molecular modeling, docking is a method which predicts the preferred orientation of one molecule to a second when bound to each other to form a stable complex. Knowledge of the preferred orientation in turn may be used to predict the strength of association or binding affinity between two molecules using for example scoring functions.The associations between biologically relevant molecules such as proteins, nucleic acids, carbohydrates, and lipids play a central role in signal transduction. Furthermore, the relative orientation of the two interacting partners may affect the type of signal produced (e.g., agonism vs antagonism). Therefore docking is useful for predicting both the strength and type of signal produced.Docking is frequently used to predict the binding orientation of small molecule drug candidates to their protein targets in order to in turn predict the affinity and activity of the small molecule. Hence docking plays an important role in the rational design of drugs. Given the biological and pharmaceutical significance of molecular docking, considerable efforts have been directed towards improving the methods used to predict docking.

Molecular docking can be thought of as a problem of "lock-and-key", where one is interested in finding the correct relative orientation of the "key" which will open up the "lock" (where on the surface of the lock is the key hole, which direction to turn the key after it is inserted, etc.). Here, the protein can be thought of as the "lock" and the ligand can be thought of as a "key". Molecular docking may be defined as an optimization problem, which would describe the "best-fit" orientation of a ligand that binds to a particular protein of interest. However since both the ligand and the protein are flexible, a "hand-in-glove" analogy is more appropriate than "lock-and-key". During the course of the process, the ligand and the protein adjust their conformation to achieve an overall "best-fit" and this kind of conformational adjustments resulting in the overall binding is referred to as "induced-fit".The focus of molecular docking is to computationally stimulate the molecular recognition process. The aim of molecular docking is to achieve an optimized conformation for both the protein and ligand and relative orientation between protein and ligand such that the free energy of the overall system is minimized.

Docking Approaches

Two approaches are particularly popular within the molecular docking community. One approach uses a matching technique that describes the protein and the ligand as complementary surfaces. The second approach simulates the actual docking process in which the ligand-protein pairwise interaction energies are calculated. Both approaches have significant advantages as well as some limitations. These are outlined below.

Shape Complementarity Methods

Geometric matching/ shape complementarity methods describe the protein and ligand as a set of features that make them dockable. These features may include molecular surface/ complementary surface descriptors. In this case, the receptor's molecular surface is described in terms of its solvent-accessible surface area and the ligand's molecular surface is described in terms of its matching surface description. The complementarity between the two surfaces amounts to the shape matching description that may help finding the complementary pose of docking the target and the ligand molecules. Another approach is to describe the hydrophobic features of the protein using turns in the main-chain atoms. Yet another approach is to use a Fourier shape descriptor technique.

Whereas the shape complementarity based approaches are typically fast and robust, they cannot usually model the movements or dynamic changes in the ligand/ protein conformations accurately, although recent developments allow these methods to investigate ligand flexibility. Shape complementarity methods can quickly scan through several thousand ligands in a matter of seconds and actually figure out whether they can bind at the protein's active site, and are usually scalable to even protein-protein interactions. They are also much more amenable to pharmacophore based approaches, since they use geometric descriptions of the ligands to find optimal binding.

Simulation Processes

The simulation of the docking process as such is a much more complicated process. In this approach, the protein and the ligand are separated by some physical distance, and the ligand finds its position into the protein's active site after a certain number of "moves" in its conformational space. The moves incorporate rigid body transformations such as translations and rotations, as well as internal changes to the ligand's structure including torsion angle rotations. Each of these moves in the conformation space of the ligand induces a total energetic cost of the system, and hence after every move the total energy of the system is calculated. The obvious advantage of the method is that it is more amenable to incorporating ligand flexibility into its modeling whereas shape complementarity techniques have to use some ingenious methods to incorporate flexibility in ligands. Another advantage is that the process is physically closer to what happens in reality, when the protein and ligand approach each other after molecular recognition. A clear disadvantage of this technique is that it takes longer time to evaluate the optimal pose of binding since they have to explore a rather large energy landscape. However grid-based techniques as well as fast optimization methods have significantly ameliorated these problems.

Mechanics of Docking

To perform a docking screen, the first requirement is a structure of the protein of interest. Usually the structure has been determined using a biophysical technique such as x-ray crystallography, or less often, NMR spectroscopy. This protein structure and a database of potential ligands serve as inputs to a docking program. The success of a docking program depends on two components: the search algorithm and the scoring function.

Searching the Conformational Space for Docking

The search space consists of all possible orientations and conformations of the protein paired with the ligand. With present computing resources, it is impossible to exhaustively explore the search space-this would involve enumerating all possible distortions of each molecule (molecules are dynamic and exist in an ensemble of conformational states) and all possible rotational and translational orientations of the ligand relative to the protein at a given level of granularity. Most docking programs in use account for a flexible ligand, and several are attempting to model a flexible protein receptor. Each "snapshot" of the pair is referred to as a pose. There are many strategies for sampling the search space. Here are some examples:

— Use a coarse-grained molecular dynamics simulation to propose energetically reasonable poses

— Use a "linear combination" of multiple structures determined for the same protein to emulate receptor flexibility

— Use a genetic algorithm to "evolve" new poses that are successively more and more likely to represent favorable binding interactions.

Scoring Function

Most scoring functions are physics-based molecular mechanics force fields that estimate the energy of the pose; a low (negative) energy indicates a stable system and thus a likely binding interaction. An alternative approach is to derive a statistical potential for interactions from a large database of protein-ligand complexes, such as the Protein Data Bank, and evaluate the fit of the pose according to this inferred potential.There are a large number of structures from X-ray crystallography for complexes between proteins and high affinity ligands, but comparatively fewer for low affinity ligands as the later complexes tend to be less stable and therefore more difficult to crystallize. Scoring functions trained with this data can dock high affinity ligands correctly, but they will also give plausible docked conformations for ligands that do not bind. This gives a large number of false positive hits, i.e., ligands predicted to bind to the protein that actually don't when placed together in a test tube.

Applications

A binding interaction between a small molecule ligand and a enzyme protein may result in activation or inhibition of the enzyme. If the protein is a receptor, ligand binding may result in agonism or antagonism. Docking is most commonly used in the field of drug design - most drugs are small organic molecules, and docking may be applied to:

— hit identification - docking combined with a scoring function can be used to quickly screen large databases of potential drugs in silico to identify molecules that are likely to bind to protein target of interest.

— lead optimization - docking can be used to predict in where and in which relative orientation a ligand binds to a protein (also referred to as the binding mode or pose). This information may in turn be used to design more potent and selective analogs.

— Bioremediation- Protein ligand docking can also be used to predict the putative pollutants that can be degradable by the biodegradative proteins.

Macromolecular Docking

Macromolecular docking is the computational modelling of the molecular structure of complexes formed by two or more interacting biological macromolecules. Protein-protein complexes are the most commonly attempted targets, followed increasingly by protein-nucleic acid complexes.The term "docking" originated in the late 1970s, with a more restricted meaning; then, "docking" meant refining a model of a complex structure by optimizing the separation between the interactors but keeping their relative orientations fixed.The biological roles of most proteins, as characterized by which other macromolecules they interact with, are known at best incompletely. Even those proteins which participate in a well-studied biological process (e.g., the Krebs cycle) may have unexpected interaction partners or functions which are unrelated to that process. Moreover, vast numbers of "hypothetical" proteins have been emerging as part of the genomic revolution of the late 1990s, proteins which, apart from their amino acid sequence, are a complete mystery.In cases of known protein-protein interactions, other questions arise. Genetic diseases are known to be caused by misfolded or mutated proteins (e.g., cystic fibrosis), and there is a desire to understand what, if any, anomalous protein-protein interactions a given mutation can cause. In the distant future, proteins may be designed to perform biological functions, and a determination of the potential interactions of such proteins will be essential.

Protein-protein docking is ultimately envisaged to address all these issues. Furthermore, since docking methods can be based on purely physical principles, even proteins of unknown function (or which have been studied relatively little) may be docked. The only prerequisite is that their molecular structure has been either determined experimentally, or can be estimated by a protein structure prediction technique.Protein-nucleic acid interactions feature prominently in the living cell. Transcription factors, which regulate gene expression, and polymerases, which catalyse replication, are composed of proteins, and the genetic material they interact with is composed of nucleic acids.

In the 1970's, complex modelling revolved around manually identifying features on the surfaces of of the interactors, and interpreting the consequences for binding, function and activity; any computer programmes were typically used at the end of the modelling process, to discriminate between the relatively few configurations which remained after all the heuristic constraints had been imposed. The first use of computers was in a study on hemoglobin interaction in sickle-cell fibres. This was followed in 1978 by work on the trypsin-BPTI complex.Computers discriminated between good and bad models using a scoring function which rewarded large interface area, and pairs of molecules in contact but not occupying the same space. The computer used a simplified representation of the interacting proteins, with one interaction centre for each residue. Favorable electrostatic interactions, including hydrogen bonds, were identified by hand.Towards the beginning of the 1990's, more structures of complexes were determined, and available computational power had increased substantially. With the emergence of bioinformatics, the focus moved towards developing generalized techniques which could be applied to an arbitrary set of complexes at acceptable computational cost. The new methods were envisaged to apply even in the absence of phylogenetic or experimental clues; any specific prior knowledge could still be introduced at the stage of choosing between the highest ranking output models, or be framed as input if the algorithm catered for it. 1992 saw the publication of the correlation method, an algorithm which used the fast Fourier transform to give a vastly improved scalability for evaluating coarse shape complementarity on rigid-body models. This was extended in 1997 to cover coarse electrostatics.

Rigid-body docking vs. flexible docking

If the bond angles, bond lengths and torsion angles of the components are not modified at any stage of complex generation, it is known as rigid body docking. A subject of speculation is whether or not rigid-body docking is sufficiently good for most docking. When substantial conformational change occurs within the components at the time of complex formation, rigid-body docking is inadequate. However, scoring all possible conformational changes is prohibitively expensive in computer time. Docking procedures which permit conformational change, or flexible docking procedures, must intelligently select small subset of possible conformational changes for consideration.

Methods

Successful docking requires two criteria:

- Generating a set configurations which reliably includes at least one nearly correct one.
- Reliably distinguishing nearly correct configurations from the others.

For many interactions, the binding site is known on one or more of the proteins to be docked. This is the case for antibodies and for competitive inhibitors. In other cases, a binding site may be strongly suggested by mutagenic or phylogenetic evidence. Configurations where the proteins interpenetrate severely may also be ruled out a priori.After making exclusions based on prior knowledge or stereochemical clash, the remaining space of possible complexed structures must be sampled exhaustively, evenly and with a sufficient coverage to guarantee a near hit. Each configuration must be scored with a measure that is capable of ranking a nearly correct structure above at least 100,000 alternatives. This is a computationally intensive task, and a variety of strategies have been developed.

Reciprocal space methods

Each of the proteins may be represented as a simple cubic lattice. Then, for the class of scores which are discrete convolutions, configurations related to each other by translation of one protein by an exact lattice vector can all be scored almost simultaneously by applying the convolution theorem. It is possible to construct reasonable, if approximate, convolution-like scoring functions representing both stereochemical and electrostatic fitness.Reciprocal space methods have been used extensively for their ability to evaluate enormous numbers of configurations. They lose their speed advantage if torsional changes are introduced. Another drawback is that it is impossible to make efficient use of prior knowledge. The question also remains whether convolutions are too limited a class of scoring function to identify the best complex reliably.

Monte Carlo methods

In Monte Carlo, an initial configuration is refined by taking random steps which are accepted or rejected based on their induced improvement in score, until a certain number of steps have been tried. The assumption is that convergence to the best structure should occur from a large class of initial configurations, only one of which needs to be considered. Initial configurations may be sampled coarsely, and much computation time can be saved. Because of the difficulty of finding a scoring function which is both highly discriminating for the correct configuration and also converges to the correct configuration from a distance, the use of two levels of refinement, with different scoring functions, has been proposed. Torsion can be introduced naturally to Monte Carlo as an additional property of each random move.Monte Carlo methods are not guaranteed to search exhaustively, so that the best configuration may be missed even using a scoring function which would in theory identify it. How severe a problem this is for docking has not been firmly established.

Selecting the docked complex structure

To find a score which forms a consistent basis for selecting the best configuration, studies are carried out on a standard benchmark of protein-protein interaction cases. Scoring functions are assessed on the rank they assign to the best structure (ideally the best structure should be ranked 1), and on their coverage (the proportion of the benchmark cases for which they achieve an acceptable result). Types of scores studied include:

- Heuristic scores based on residue contacts.
- Shape complementarity of molecular surfaces ("stereochemistry").
- Free energies, estimated using parameters from molecular mechanics force fields such as CHARMM or AMBER.
- Phylogenetic desirability of the interacting regions.

— Clustering coefficients.

It is usual to create hybrid scores by combining one or more categories above in a weighted sum whose weights are optimized on cases from the benchmark. To avoid bias, the benchmark cases used to optimize the weights must not overlap with the cases used to make the final test of the score.A benchmark of 84 protein-protein interactions with known complexed structures has been developed for testing docking methods. The set is chosen to cover a wide range of interaction types, and to avoid repeated features, such as the profile of interactors' structural families according to the SCOP database. Benchmark elements are classified into three levels of difficulty (the most difficult containing the largest change in backbone conformation). The protein-protein docking benchmark contains examples of enzyme-inhibitor, antigen-antibody and homomultimeric complexes.

CAPRI assessment

The Critical Assessment of PRediction of Interactions is an ongoing series of events in which researchers throughout the community try to dock the same proteins, as provided by the assessors. Rounds take place approximately every 6 months. Each round contains between one and six target protein-protein complexes whose structures have been recently determined experimentally. The coordinates and are held privately by the assessors, with the cooperation of the structural biologists who determined them. The assessment of submissions is double blind.CAPRI attracts a high level of participation (37 groups participated worldwide in round seven) and a high level of interest from the biological community in general. Although CAPRI results are of little statistical significance owing to the small number of targets in each round, the role of CAPRI in stimulating discourse is significant. (The CASP assessment is a similar exercise in the field of protein structure prediction).A reliable method for affinity prediction has the potential to transform biochemistry and cell biology. Though a distant prospect, affinity prediction may be considered as the ultimate achievement in protein-protein docking.

Protein-protein docking and molecular docking

The field of protein-protein docking is highly computationally oriented, and it shares approaches with small-molecule docking. Proteins complexed with polynucleotide molecules are widely studied using similar or identical approaches to protein-protein docking, although if the nucleotide molecule is small enough, the case may be framed as a small-molecule docking problem.

CPK coloring

CPK Coloring is a way to color atoms when visualizing molecular models, inspired by CPK plastic molecular model kits by Corey, Pauling and Koltun. The method uses the van der Waals radii to visualize atoms.

CPK coloring conventionally depicts:

- hydrogen as white,
- carbon as black or grey,
- nitrogen as blue,
- oxygen as red,
- phosphorus as orange, and
- sulfur as yellow.

Other colors that are not always standard include:

- green for magnesium and / or chlorine,
- orange for iron, and
- pink for unknown elements or those with atomic numbers over 57.

Scoring functions for docking

In the fields of computational chemistry and molecular modelling, scoring functions are fast approximate mathematical methods used to predict the strength of the non-covalent interaction (also referred to as binding affinity) between two molecules after they have been docked. Most commonly one of the molecules is a small organic compound such as a drug and the second is the drug's biological target such as a protein receptor. Scoring functions have also been developed to predict the strength of other types of intermolecular interactions, for example between two proteins or between protein and DNA.

Utility

Scoring functions are particularly useful for drug and other types of molecular design where speed is more important than high accuracy. These application include:

- Virtual screening of small molecule databases of candidate ligands to identify novel small molecules that bind to a protein target of interest and therefore are useful starting points for drug discovery
- De novo design (design "from scratch") of novel small molecules that bind to a protein target
- Lead optimization of screening hits to optimize their affinity and selectivity

A potentially more reliable but much more computationally demanding alternative to scoring functions are free energy perturbation calculations.

Prerequisites

Scoring functions are normally parameterized (or trained) against a data set consisting of experimentally determined binding affinities between molecular species similar to the species that one wishes to predict.

For predictions of affinities of ligands for proteins the following must first be known or predicted:

— Protein tertiary structure - arrangement of the protein atoms in three dimensional space. Protein structures may be determined by experimental techniques such as X-ray crystallography or solution phase NMR methods or predicted by homology modelling.

— Ligand active conformation - three dimensional shape of the ligand when bound to the protein

— Binding-mode - orientation of the two binding partners relative to each other in the complex

The above information yields the three dimensional structure of the complex. Based on this structure, the scoring function can then estimate the strength of the association between the two molecules in the complex using one of the methods outlined below. Finally the scoring function itself may be used to help predict both the binding mode and the active conformation of the small molecule in the complex.

Classes

There are three general classes of scoring functions:

— Force field - affinities are estimated by summing the strength of intermolecular van der Waals and electrostatic interactions between all atoms of the two molecules in the complex. The intramolecular energies (also referred to as strain energy) of the two binding partners are also frequently included. Finally since the binding normally takes place in the presence of water, the desolvation energies of the ligand and of the protein are sometimes taken into account using implicit solvation methods such as GBSA or PBSA.

— Empirical - based on counting the number off various types of interactions between the two binding partners. Counting may be based on the number of ligand and receptor atoms in contact with each other or by calculating the change in solvent accessible surface area (DSASA) in the complex compared to the uncomplexed ligand and protein. The coefficients of the scoring function are usually fit using multiple linear regression methods. These interactions terms of the function may include for example:

  - number of hydrogen bonds (favorable electrostatic contribution to affinity),
  - hydrophobic - hydrophobic contacts (favorable),
  - hydrophilic-hydrophobic contacts (unfavorable),
  - number of rotatable bonds immobilized in complex formation (unfavorable entropic contribution).
- Knowledge - based on statistical observations of intermolecular close contacts in large 3D databases (such as the Cambridge Structural Database or Protein Data Bank) which are used to derive "potentials of mean force". This method is founded on the assumption that close intermolecular interactions between certain types of atoms or functional groups that occur more frequently than one would expect by a random distribution are likely to be energetically favorable and therefore contribute favorably to binding affinity.

Finally hybrid scoring functions have also been developed in which the components from two or more of the above scoring functions are combined into one function.

PROTEIN-LIGAND DOCKING

Protein-ligand docking is a molecular modelling technique. The goal of protein-ligand docking is to predict the position and orientation of a ligand (a small molecule) when it is bound to a protein receptor or enzyme. Pharmaceutical research employs docking techniques for a variety of purposes, most notably in the virtual screening of large databases of available chemicals in order to select likely drug candidates.Several protein-ligand docking softwares are available, such as Autodock or EADock. There is a web service Molecular Docking Server that calculates the site, geometry and energy of small molecules interacting with proteins.

MOLECULAR DYNAMICS

Molecular dynamics (MD) is a form of computer simulation in which atoms and molecules are allowed to interact for a period of time by approximations of known physics, giving a view of the motion of the atoms. Because molecular systems generally consist of a vast number of particles, it is impossible to find the properties of such complex systems analytically; MD simulation circumvents this problem by using numerical methods. It represents an interface between laboratory experiments and theory, and can be understood as a "virtual experiment". MD probes the relationship between molecular structure, movement and function. Molecular dynamics is a multidisciplinary method. Its laws and theories stem from mathematics, physics, and chemistry, and it employs algorithms from computer science and information theory. It was originally conceived within theoretical physics in the late 1950s, but is applied today mostly in materials science and biomolecules.

Before it became possible to simulate molecular dynamics with computers, some undertook the hard work of trying it with physical models such as macroscopic spheres. The idea was to arrange them to replicate the properties of a liquid. J.D. Bernal said, in 1962: "... I took a number of rubber balls and stuck them together with rods of a selection of different lengths ranging from 2.75 to 4 inches. I tried to do this in the first place as casually as possible, working in my own office, being interrupted every five minutes or so and not remembering what I had done before the interruption." Fortunately, now computers keep track of bonds during a simulation.

Molecular dynamics is a specialized discipline of molecular modeling and computer simulation based on statistical mechanics; the main justification of the MD method is that statistical ensemble averages are equal to time averages of the system, known as the ergodic hypothesis. MD has also been termed "statistical mechanics by numbers" and "Laplace's vision of Newtonian mechanics" of predicting the future by animating nature's forces and allowing insight into molecular motion on an atomic scale. However, long MD simulations are mathematically ill-conditioned, generating cumulative errors in numerical integration that can be minimized with proper selection of algorithms and parameters, but not eliminated entirely. Furthermore, current potential functions are, in many cases, not sufficiently accurate to reproduce the dynamics of molecular systems, so the much more computationally demanding Ab Initio Molecular Dynamics method must be used. Nevertheless, molecular dynamics techniques allow detailed time and space resolution into representative behavior in phase space.

Application

There is a significant difference between the focus and methods used by chemists and physicists, and this is reflected in differences in the jargon used by the different fields. In chemistry and biophysics, the interaction between the particles is either described by a "force field" (classical MD), a quantum chemical model, or a mix between the two. These terms are not used in physics, where the interactions are usually described by the name of the theory or approximation being used and called the potential energy, or just "potential".Beginning in theoretical physics, the method of MD gained popularity in materials science and since the 1970s also in biochemistry and biophysics. In chemistry, MD serves as an important tool in protein structure determination and refinement using experimental tools such as X-ray crystallography and NMR. It has also been applied with limited success as a method of refining protein structure predictions. In physics, MD is used to examine the dynamics of atomic-level phenomena that cannot be observed directly, such as thin film growth and ion-subplantation. It is also used to examine the physical properties of nanotechnological devices that have not or cannot yet be created.

In applied mathematics and theoretical physics, molecular dynamics is a part of the research realm of dynamical systems, ergodic theory and statistical mechanics in general.

The concepts of energy conservation and molecular entropy come from thermodynamics. Some techniques to calculate conformational entropy such as principal components analysis come from information theory. Mathematical techniques such as the transfer operator become applicable when MD is seen as a Markov chain. Also, there is a large community of mathematicians working on volume preserving, symplectic integrators for more computationally efficient MD simulations.

Design Constraints

Design of a molecular dynamics simulation should account for the available computational power. Simulation size (n=number of particles), timestep and total time duration must be selected so that the calculation can finish within a reasonable time period. However, the simulations should be long enough to be relevant to the time scales of the natural processes being studied. To make statistically valid conclusions from the simulations, the time span simulated should match the kinetics of the natural process. Otherwise, it is analogous to making conclusions about how a human walks from less than one footstep. Most scientific publications about the dynamics of proteins and DNA use data from simulations spanning nanoseconds (1E-9 s) to microseconds (1E-6 s). To obtain these simulations, several CPU-days to CPU-years are needed. Parallel algorithms allow the load to be distributed among CPUs; an example is the spatial decomposition in LAMMPS.

During a classical MD simulation, the most CPU intensive task is the evaluation of the potential (force field) as a function of the particles' internal coordinates. Within that energy evaluation, the most expensive one is the non-bonded or non-covalent part. In Big O notation, common molecular dynamics simulations scale by $O(n^2)$ if all pair-wise electrostatic and van der Waals interactions must be accounted for explicitly. This computational cost can be reduced by employing electrostatics methods such as Particle Mesh Ewald ($O(n\log(n))$) or good spherical cutoff techniques ($O(n)$).Another factor that impacts total CPU time required by a simulation is the size of the integration timestep. This is the time length between evaluations of the potential. The timestep must be chosen small enough to avoid discretization errors (i.e. smaller than the fastest vibrational frequency in the system). Typical timesteps for classical MD are in the order of 1 femtosecond (1E-15 s). This value may be extended by using algorithms such as SHAKE, which fix the vibrations of the fastest atoms (e.g. hydrogens) into place. Multiple time scale methods have also been developed, which allow for extended times between updates of slower long-range forces.

For simulating molecules in a solvent, a choice should be made between explicit solvent and implicit solvent. Explicit solvent particles (such as the TIP3P and SPC/E water models) must be calculated expensively by the force field, while implicit solvents use a mean-field approach. The impact of explicit solvents on CPU-time can be 10-fold or more. But the granularity and viscosity of explicit solvent is essential to reproduce certain properties of the solute molecules.In all kinds of molecular dynamics simulations, the simulation box size must be large enough to avoid boundary condition artifacts. Boundary conditions are often treated by choosing fixed values at the edges, or by employing periodic boundary conditions in which one side of the simulation loops back to the opposite side, mimicking a bulk phase.

Microcanonical Ensemble (NVE)

In the microcanonical, or NVE ensemble, the system is isolated from changes in moles (N), volume (V) and energy (E). It corresponds to an adiabatic process with no heat exchange. A microcanonical molecular dynamics trajectory may be seen as an exchange of potential and kinetic energy, with total energy being conserved. For a system of N particles with coordinates X and velocities V, the following pair of first order differential equations may be written in Newton's notation as

$$F(X) = -\nabla V(X) = MV(t)$$

$$V(t) = X(t)$$

The potential energy function $U(X)$ of the system is a function of the particle coordinates X. It is referred to simply as the "potential" in Physics, or the "force field" in Chemistry. The first equation comes from Newton's laws; the force F acting on each particle in the system can be calculated as the negative gradient of $U(X)$.For every timestep, each particle's position X and velocity V may be integrated with a symplectic method such as Verlet. The time evolution of X and V is called a trajectory. Given the initial positions (e.g. from theoretical knowledge) and velocities (e.g. randomized Gaussian), we can calculate all future (or past) positions and velocities.

One frequent source of confusion is the meaning of temperature in MD. Commonly we have experience with macroscopic temperatures, which involve a huge number of particles. But temperature is a statistical quantity. If there is a large enough number of atoms, statistical temperature can be estimated from the instantaneous temperature, which is found by equating the kinetic energy of the system to $nk_BT/2$ where n is the number of degrees of freedom of the system.A temperature-related phenomenon arises due to the small number of atoms that are used in MD simulations. For example, consider simulating the growth of a copper film starting with a substrate containing 500 atoms and a deposition energy of 100 eV. In the real world, the 100 eV from the deposited atom would rapidly be transported through and shared among a large number of atoms (10^{10} or more) with no big change in temperature. When there are only 500 atoms, however, the substrate is almost immediately vaporized by the deposition. Something similar happens in biophysical simulations. The temperature of the system in NVE is naturally raised when macromolecules such as proteins undergo exothermic conformational changes and binding.

Canonical Ensemble (NVT)

In the canonical ensemble, moles (N), volume (V) and temperature (T) are conserved. It is also sometimes called constant temperature molecular dynamics (CTMD). In NVT, the energy of endothermic and exothermic processes is exchanged with a thermostat.A variety of thermostat methods are available to add and remove energy from the boundaries of an MD system in a realistic way, approximating the canonical ensemble. Popular techniques to control temperature include the Nosé-Hoover thermostat and Langevin dynamics.

Isothermal-Isobaric (NPT) Ensemble

In the isothermal-isobaric ensemble, moles (N), pressure (P) and temperature (T) are conserved. In addition to a thermostat, a barostat is needed. It corresponds most closely to laboratory conditions with a flask open to ambient temperature and pressure.In the simulation of biological membranes, isotropic pressure control is not appropriate. For lipid bilayers, pressure control occurs under constant membrane area (NPAT) or constant surface tension "gamma" (NPγT).

Generalized Ensembles

The replica exchange method is a generalized ensemble. It was originally created to deal with the slow dynamics of disordered spin systems. It is also called parallel tempering. The replica exchange MD (REMD) formulation tries to overcome the multiple-minima problem by exchanging the temperature of non-interacting replicas of the system running at several temperatures.

Potentials in MD Simulations

A molecular dynamics simulation requires the definition of a potential function, or a description of the terms by which the particles in the simulation will interact. In chemistry and biology this is usually referred to as a force field. Potentials may be defined at many levels of physical accuracy; those most commonly used in chemistry are based on molecular mechanics and embody a classical treatment of particle-particle interactions that can reproduce structural and conformational changes but usually cannot reproduce chemical reactions.The reduction from a fully quantum description to a classical potential entails two main approximations. The first one is the Born-Oppenheimer approximation, which states that the dynamics of electrons is so fast that they can be considered to react instantaneously to the motion of their nuclei. As a consequence, they may be treated separately. The second one treats the nuclei, which are much heavier than electrons, as point particles that follow classical Newtonian dynamics. In classical molecular dynamics the effect of the electrons is approximated as a single potential energy surface, usually representing the ground state.When finer levels of detail are required, potentials based on quantum mechanics are used; some techniques attempt to create hybrid classical/ quantum potentials where the bulk of the system is treated classically but a small region is treated as a quantum system, usually undergoing a chemical transformation.

Empirical Potentials

Empirical potentials used in chemistry are frequently called force fields, while those used in materials physics are called just empirical or analytical potentials. Most force fields in chemistry are empirical and consist of a summation of bonded forces associated with chemical bonds, bond angles, and bond dihedrals, and non-bonded forces associated with van der Waals forces and electrostatic charge. Empirical potentials represent quantum-mechanical effects in a limited way through ad-hoc functional approximations. These potentials contain free parameters such as atomic charge, van der Waals parameters reflecting estimates of atomic radius, and equilibrium bond length, angle, and dihedral; these are obtained by fitting against detailed electronic calculations (quantum chemical simulations) or experimental physical properties such as elastic constants, lattice parameters and spectroscopic measurements.

Because of the non-local nature of non-bonded interactions, they involve at least weak interactions between all particles in the system. Its calculation is normally the bottleneck in the speed of MD simulations. To lower the computational cost, force fields employ numerical approximations such as shifted cutoff radii, reaction field algorithms, particle mesh Ewald summation, or the newer Particle-Particle Particle Mesh (P3M).Chemistry force fields commonly employ preset bonding arrangements (an exception being ab-initio dynamics), and thus are unable to model the process of chemical bond breaking and reactions explicitly. On the other hand, many of the potentials used in physics, such as those based on the bond order formalism can describe several different coordinations of a system and bond breaking. Examples of such potentials include the Brenner potential for hydrocarbons and its further developments for the C-Si-H and C-O-H systems. The ReaxFF potential can be considered a fully reactive hybrid between bond order potentials and chemistry force fields.

Pair potentials vs. many-body potentials

The potential functions representing the non-bonded energy are formulated as a sum over interactions between the particles of the system. The simplest choice, employed in many popular force fields, is the "pair potential", in which the total potential energy can be calculated from the sum of energy contributions between pairs of atoms. In many-body potentials, the potential energy includes the effects of three or more particles interacting with each other. In simulations with pairwise potentials, global interactions in the system also exist, but they occur only through pairwise terms. In many-body potentials, the potential energy cannot be found by a sum over pairs of atoms, as these interactions are calculated explicitly as a combination of higher-order terms. In the statistical view, the dependency between the variables cannot in general be expressed using only pairwise products of the degrees of freedom. For example, the Tersoff potential, which was originally used to simulate carbon, silicon and germanium and has since been used for a wide range of other materials, involves a sum over groups of three atoms, with the angles between the atoms being an important factor in the potential. Other examples are the embedded-atom method (EAM) and the Tight-Binding Second Moment Approximation (TBSMA) potentials, where the electron density of states in the region of an atom is calculated from a sum of contributions from surrounding atoms, and the potential energy contribution is then a function of this sum.

Semi-empirical potentials

Semi-empirical potentials make use of the matrix representation from quantum mechanics. However, the values of the matrix elements are found through empirical formulae that estimate the degree of overlap of specific atomic orbitals. The matrix is then diagonalized to determine the occupancy of the different atomic orbitals, and empirical formulae are used once again to determine the energy contributions of the orbitals.There are a wide variety of semi-empirical potentials, known as tight-binding potentials, which vary according to the atoms being modeled.

Polarizable potentials

Most classical force fields implicitly include the effect of polarizability, e.g. by scaling up the partial charges obtained from quantum chemical calculations. These partial charges are stationary with respect to the mass of the atom. But molecular dynamics simulations can explicitly model polarizability with the introduction of induced dipoles through different methods, such as Drude particles or fluctuating charges. This allows for a dynamic redistribution of charge between atoms which responds to the local chemical environment.For many years, polarizable MD simulations have been touted as the next generation. For homogenous liquids such as water, increased accuracy has been achieved through the inclusion of polarizability. Some promising results have also been achieved for proteins. However, it is still uncertain how to best approximate polarizability in a simulation.

Ab-initio methods

In classical molecular dynamics, a single potential energy surface (usually the ground state) is represented in the force field. This is a consequence of the Born-Oppenheimer approximation. If excited states, chemical reactions or a more accurate representation is needed, electronic behavior can be obtained from first principles by using a quantum mechanical method, such as Density Functional Theory. This is known as Ab Initio Molecular Dynamics (AIMD). Due to the cost of treating the electronic degrees of freedom, the computational cost of this simulations is much higher than classical molecular dynamics. This implies that AIMD is limited to smaller systems and shorter periods of time.Ab-initio quantum-mechanical methods may be used to calculate the potential energy of a system on the fly, as needed for conformations in a trajectory. This calculation is usually made in the close neighborhood of the reaction coordinate. Although various approximations may be used, these are based on theoretical considerations, not on empirical fitting. Ab-Initio calculations produce a vast amount of information that is not available from empirical methods, such as density of electronic states or other electronic properties. A significant advantage of using ab-initio methods is the ability to study reactions that involve breaking or formation of covalent bonds, which correspond to multiple electronic states.

Hybrid QM/MM

QM (quantum-mechanical) methods are very powerful however they are computationally expensive, while the MM (classical or molecular mechanics) methods are fast but suffer from several limitations (require extensive parameterization; energy estimates obtained are not very accurate; cannot be used to simulate reactions where covalent bonds are broken/formed; and are limited in their abilities for providing accurate details regarding the chemical environment). A new class of method has emerged that combines the good points of QM (accuracy) and MM (speed) calculations. These methods are known as mixed or hybrid quantum-mechanical and molecular mechanics methods (hybrid QM/MM). The methodology for such techniques was introduced by Warshel and coworkers. In the recent years have been pioneered by several groups including: Arieh Warshel (University of Southern California), Weitao Yang (Duke University), Sharon Hammes-Schiffer (The Pennsylvania State University), Donald Truhlar and Jiali Gao (University of Minnesota) and Kenneth Merz (University of Florida).

The most important advantage of hybrid QM/MM methods is the speed. The cost of doing classical molecular dynamics (MM) in the most straightforward case scales $O(n^2)$, where N is the number of atoms in the system. This is mainly due to electrostatic interactions term (every particle interacts with every other particle). However, use of cutoff radius, periodic pair-list updates and more recently the variations of the particle-mesh Ewald's (PME) method has reduced this between $O(N)$ to $O(n^2)$. In other words, if a system with twice many atoms is simulated then it would take between twice to four times as much computing power. On the other hand the simplest ab-initio calculations typically scale $O(n^3)$ or worse (Restricted Hartree-Fock calculations have been suggested to scale $\sim O(n^{2.7})$). To overcome the limitation, a small part of the system is treated quantum-mechanically (typically active-site of an enzyme) and the remaining system is treated classically.In more sophisticated implementations, QM/MM methods exist to treat both light nuclei susceptible to quantum effects (such as hydrogens) and electronic states. This allows generation of hydrogen wave-functions (similar to electronic wave-functions). This methodology has been useful in investigating phenomenon such as hydrogen tunneling. One example where QM/MM methods have provided new discoveries is the calculation of hydride transfer in the enzyme liver alcohol dehydrogenase. In this case, tunneling is important for the hydrogen, as it determines the reaction rate.

Coarse-graining and reduced representations

At the other end of the detail scale are coarse-grained and lattice models. Instead of explicitly representing every atom of the system, one uses "pseudo-atoms" to represent groups of atoms. MD simulations on very large systems may require such large computer resources that they cannot easily be studied by traditional all-atom methods. Similarly, simulations of processes on long timescales (beyond about 1 microsecond) are prohibitively expensive, because they require so many timesteps. In these cases, one can sometimes tackle the problem by using reduced representations, which are also called coarse-grained models.Examples for coarse graining (CG) methods are discontinuous molecular dynamics (CG-DMD) and Go-models. Coarse-graining is done sometimes taking larger pseudo-atoms. Such united atom approximations have been used in MD simulations of biological membranes. The aliphatic tails of lipids are represented by a few pseudo-atoms by gathering 2-4 methylene groups into each pseudo-atom.

The parameterization of these very coarse-grained models must be done empirically, by matching the behavior of the model to appropriate experimental data or all-atom simulations. Ideally, these parameters should account for both enthalpic and entropic contributions to free energy in an implicit way. When coarse-graining is done at higher levels, the accuracy of the dynamic description may be less reliable. But very coarse-grained models have been used successfully to examine a wide range of questions in structural biology.

Examples of applications of coarse-graining in biophysics:

- protein folding studies are often carried out using a single (or a few) pseudo-atoms per amino acid;
- DNA supercoiling has been investigated using 1-3 pseudo-atoms per basepair, and at even lower resolution;
- Packaging of double-helical DNA into bacteriophage has been investigated with models where one pseudo-atom represents one turn (about 10 basepairs) of the double helix;
- RNA structure in the ribosome and other large systems has been modeled with one pseudo-atom per nucleotide.

The simplest form of coarse-graining is the "united atom" (sometimes called "extended atom") and was used in most early MD simulations of proteins, lipids and nucleic acids. For example, instead of treating all four atoms of a CH_3 methyl group explicitly (or all three atoms of CH_2 methylene group), one represents the whole group with a single pseudo-atom. This pseudo-atom must, of course, be properly parameterized so that its van der Waals interactions with other groups have the proper distance-dependence. Similar considerations apply to the bonds, angles, and torsions in which the pseudo-atom participates. In this kind of united atom representation, one typically eliminates all explicit hydrogen atoms except those that have the capability to participate in hydrogen bonds ("polar hydrogens"). The polar hydrogens are usually retained in the model, because proper treatment of hydrogen bonds requires a reasonably accurate description of the directionality and the electrostatic interactions between the donor and acceptor groups. A hydroxyl group, for example, can be both a hydrogen bond donor and a hydrogen bond acceptor, and it would be impossible to treat this with a single OH pseudo-atom. Note that about half the atoms in a protein or nucleic acid are nonpolar hydrogens, so the use of united atoms can provide a substantial savings in computer time.

REFERENCES

Barlow, W., "Probable Nature of the Internal Symmetry of Crystals", *Nature*, 1883.

Cai W, Shao X, Maigret B., "Protein-ligand recognition using spherical harmonic molecular surfaces: towards a fast and efficient filter for large virtual throughput screening", *J. Mol. Graph. Model.*, January 2002.

Dickerson, R.E.; Geis, I., *The structure and action of proteins*, Menlo Park, CA: W.A. Benjamin, 1969.

Drexler, K. E., "Molecular Engineering: an approach to the development of general capabilities for molecular manipulation," *Proc. Nat. Acad. Sci.* (USA), Vol 78, 1981, pp. 5275–5278

Feynman, R., "There's Plenty of Room at the Bottom," in: Gilbert, H. D. (ed.), *Miniaturization*, Reinhold, New York, 1961, pp. 282–296

Kinloch A.J., *Adhesion and Adhesives: Science and Technology*, Chapman and Hall, 1987.

Morris R.J, Najmanovich R.J, Kahraman A, Thornton J.M., "Real spherical harmonic expansion coefficients as 3D shape descriptors for protein binding pocket and ligand comparisons", *Bioinformatics*, May 2005.

Noel S. Hush, "An Overview of the First Half-Century of Molecular Electronics", *Ann. N.Y. Acad. Sci.*, 2003.

Rastetter, W. H. "Enzyme Engineering," *Applied Biochem. and Biotech.*, Vol. 8, 1983, pp.423–436

Weiner, S. J., et al., "A New Force Field for Molecular Mechanical Simulation of Nucleic Acids and Proteins," *J. Am. Chem. Soc.*, Vol. 106, 1984, pp. 765–784.

5

Molecular Biology of Evolution

Evolution is change in the genetic material of a population of organisms from one generation to the next. Though changes produced in any one generation are small, differences accumulate with each generation and can, over time, cause substantial changes in the population, a process that can culminate in the emergence of new species. Indeed, the similarities amongst species suggest that all known species are descended from a common ancestor (or ancestral gene pool) through this process of gradual divergence.The basis of evolution is the genes that are passed on from generation to generation; these produce an organism's inherited traits. These traits vary within populations, with organisms showing heritable differences (variation) in their traits. Evolution itself is the product of two opposing forces: processes that constantly introduce variation, and processes that make variants either become more common or rare. New variation arises in two main ways: either from mutations in genes, or from the transfer of genes between populations and between species. In species that reproduce sexually, new combinations of genes are also produced by genetic recombination, which can increase variation between organisms.

Two major mechanisms determine which variants will become more common or rare in a population. One is natural selection, a process that causes helpful traits (those that increase the chance of survival and reproduction) to become more common in a population and causes harmful traits to become more rare. This occurs because individuals with advantageous traits are more likely to reproduce, meaning that more individuals in the next generation will inherit these traits. Over many generations, adaptations occur through a combination of successive, small, random changes in traits, and natural selection of the variants best-suited for their environment. The other major mechanism driving evolution is genetic drift, an independent process that produces random changes in the frequency of traits in a population. Genetic drift results from the role that chance plays in whether a given trait will be passed on as individuals survive and reproduce.

Evolution in organisms occurs through changes in heritable traits—particular characteristics of an organism. In humans, for example, eye color is an inherited characteristic, which individuals can inherit from one of their parents. Inherited traits are controlled by genes and the complete set of genes within an organism's genome is called its genotype. The complete set of observable traits that make up the structure and behavior of an organism is called its phenotype. These traits come from the interaction of its genotype with the environment. As a result, not every aspect of an organism's phenotype is inherited. Suntanned skin results from the interaction between a person's genotype and sunlight; thus, suntans are not passed on to people's children. However, people have different responses to sunlight, arising from differences in their genotype; a striking example is individuals with the inherited trait of albinism, who do not tan and are highly sensitive to sunburn.

Heritable traits are passed from one generation to the next via DNA, a molecule that encodes genetic information. DNA is a polymer composed of four types of bases. The sequence of bases along a particular DNA molecule specify the genetic information, in a manner similar to a sequence of letters specifying a sentence. Portions of a DNA molecule that specify a single functional unit are called genes; different genes have different sequences of bases. Within cells, the long strands of DNA form condensed structures called chromosomes. A specific location within a chromosome is known as a locus. If the DNA sequence at a locus varies between individuals, the different forms of this sequence are called alleles. DNA sequences can change through mutations, producing new alleles. If a mutation occurs within a gene, the new allele may affect the trait that the gene controls, altering the phenotype of the organism.However, while this simple correspondence between an allele and a trait works in some cases, most traits are more complex and are controlled by multiple interacting genes. The study of such complex traits is a major area of current genetic research. Another interesting but unsolved question in genetics is if epigenetics is important in evolution, this is where heritable changes occur in organisms without there being any changes to the sequences of their genes.

GENETIC VARIATION

Evolution can be divided into microevolution and macroevolution. The kind of evolution documented above is microevolution. Larger changes, such as when a new species is formed, are called macroevolution. Some biologists feel the mechanisms of macroevolution are different from those of microevolutionary change. Others think the distinction between the two is arbitrary — macroevolution is cumulative microevolution. The word evolution has a variety of meanings. The fact that all organisms are linked via descent to a common ancestor is often called evolution. The theory of how the first living organisms appeared is often called evolution. This should be called abiogenesis. And frequently, people use the word evolution when they really mean natural selection — one of the many mechanisms of evolution.

Evolution requires genetic variation. If there were no dark moths, the population could not have evolved from mostly light to mostly dark. In order for continuing evolution there must be mechanisms to increase or create genetic variation and mechanisms to decrease it. Mutation is a change in a gene. These changes are the source of new genetic variation. Natural selection operates on this variation. Genetic variation has two components: allelic diversity and non- random associations of alleles. Alleles are different versions of the same gene. For example, humans can have A, B or O alleles that determine one aspect of their blood type. Most animals, including humans, are diploid — they contain two alleles for every gene at every locus, one inherited from their mother and one inherited from their father. Locus is the location of a gene on a chromosome. Humans can be AA, AB, AO, BB, BO or OO at the blood group locus. If the two alleles at a locus are the same type (for instance two A alleles) the individual would be called homozygous. An individual with two different alleles at a locus (for example, an AB individual) is called heterozygous. At any locus there can be many different alleles in a population, more alleles than any single organism can possess. For example, no single human can have an A, B and an O allele. Considerable variation is present in natural populations. At 45 percent of loci in plants there is more than one allele in the gene pool. Any given plant is likely to be heterozygous at about 15 percent of its loci. Levels of genetic variation in animals range from roughly 15% of loci having more than one allele (polymorphic) in birds, to over 50% of loci being polymorphic in insects. Mammals and reptiles are polymorphic at about 20% of their loci—amphibians and fish are polymorphic at around 30% of their loci. In most populations, there are enough loci and enough different alleles that every individual, identical twins excepted, has a unique combination of alleles.

Linkage disequilibrium is a measure of association between alleles of two different genes. If two alleles were found together in organisms more often than would be expected, the alleles are in linkage disequilibrium. If there two loci in an organism (A and B) and two alleles at each of these loci (A1, A2, B1 and B2) linkage disequilibrium (D) is calculated as D = f(A1B1) * f(A2B2) - f(A1B2) * f(A2B1) (where f(X) is the frequency of X in the population). D varies between -1/4 and 1/4; the greater the deviation from zero, the greater the linkage. The sign is simply a consequence of how the alleles are numbered. Linkage disequilibrium can be the result of physical proximity of the genes. Or, it can be maintained by natural selection if some combinations of alleles work better as a team.

Natural selection maintains the linkage disequilibrium between color and pattern alleles in *Papilio memnon*. In this moth species, there is a gene that determines wing morphology. One allele at this locus leads to a moth that has a tail; the other allele codes for a untailed moth. There is another gene that determines if the wing is brightly or darkly colored. There are thus four possible types of moths: brightly colored moths with and without tails, and dark moths with and without tails. All four can be produced when moths are brought into the lab and bred. However, only two of these types of moths are found in the wild: brightly colored moths with tails and darkly colored moths without tails. The non-random association is maintained by natural selection. Bright, tailed moths mimic the pattern of an unpalatable species. The dark morph is cryptic. The other two combinations are neither mimetic nor cryptic and are quickly eaten by birds. Assortative mating causes a non-random distribution of alleles at a single locus. If there are two alleles (A and a) at a locus with frequencies p and q, the frequency of the three possible genotypes (AA, Aa and aa) will be p2, 2pq and q2, respectively. For example, if the frequency of A is 0.9 and the frequency of a is 0.1, the frequencies of AA, Aa and aa individuals are: 0.81, 0.18 and 0.01. This distribution is called the Hardy-Weinberg equilibrium. Non-random mating results in a deviation from the Hardy-Weinberg distribution. Humans mate assortatively according to race; we are more likely to mate with someone of own race than another. In populations that mate this way, fewer heterozygotes are found than would be predicted under random mating. A decrease in heterozygotes can be the result of mate choice, or simply the result of population subdivision. Most organisms have a limited dispersal capability, so their mate will be chosen from the local population.

Genetic Variation Mechanisms

Natural Selection

Some types of organisms within a population leave more offspring than others. Over time, the frequency of the more prolific type will increase. The difference in reproductive capability is called natural selection. Natural selection is the only mechanism of adaptive evolution; it is defined as differential reproductive success of pre- existing classes of genetic variants in the gene pool. The most common action of natural selection is to remove unfit variants as they arise via mutation. In other words, natural selection usually prevents new alleles from increasing in frequency. This led a famous evolutionist, George Williams, to say "Evolution proceeds in spite of natural selection."

Natural selection can maintain or deplete genetic variation depending on how it acts. When selection acts to weed out deleterious alleles, or causes an allele to sweep to fixation, it depletes genetic variation. When heterozygotes are more fit than either of the homozygotes, however, selection causes genetic variation to be maintained. This is called balancing selection. An example of this is the maintenance of sickle-cell alleles in human populations subject to malaria. Variation at a single locus determines whether red blood cells are shaped normally or sickled. If a human has two alleles for sickle-cell, he/she develops anemia — the shape of sickle-cells precludes them carrying normal levels of oxygen. However, heterozygotes who have one copy of the sickle-cell allele, coupled with one normal allele enjoy some resistance to malaria — the shape of sickled cells make it harder for the plasmodia (malaria causing agents) to enter the cell. Thus, individuals homozygous for the normal allele suffer more malaria than heterozygotes. Individuals homozygous for the sickle- cell are anemic. Heterozygotes have the highest fitness of these three types. Heterozygotes pass on both sickle-cell and normal alleles to the next generation. Thus, neither allele can be eliminated from the gene pool. The sickle-cell allele is at its highest frequency in regions of Africa where malaria is most pervasive.

Balancing selection is rare in natural populations. Only a handful of other cases beside the sickle-cell example have been found. At one time population geneticists thought balancing selection could be a general explanation for the levels of genetic variation found in natural populations. That is no longer the case. Balancing selection is only rarely found in natural populations. And, there are theoretical reasons why natural selection cannot maintain polymorphisms at several loci via balancing selection. Individuals are selected. Dark colored moths had a higher reproductive success because light colored moths suffered a higher predation rate. The decline of light colored alleles was caused by light colored individuals being removed from the gene pool. Individual organisms either reproduce or fail to reproduce and are hence the unit of selection. One way alleles can change in frequency is to be housed in organisms with different reproductive rates. Genes are not the unit of selection; neither are groups of organisms a unit of selection. There are some exceptions to this "rule," but it is a good generalisation.

Organisms do not perform any behaviors that are for the good of their species. An individual organism competes primarily with others of it own species for its reproductive success. Natural selection favors selfish behavior because any truly altruistic act increases the recipient's reproductive success while lowering the donors. Altruists would disappear from a population as the non- altruists would reap the benefits, but not pay the costs, of altruistic acts. Many behaviors appear altruistic. Biologists, however, can demonstrate that these behaviors are only apparently altruistic. Cooperating with or helping other organisms is often the most selfish strategy for an animal. This is called reciprocal altruism. A good example of this is blood sharing in vampire bats. In these bats, those lucky enough to find a meal will often share part of it with an unsuccessful bat by regurgitating some blood into the other's mouth. Biologists have found that these bats form bonds with partners and help each other out when the other is needy. If a bat is found to be a "cheater," (he accepts blood when starving, but does not donate when his partner is) his partner will abandon him. The bats are thus not helping each other altruistically; they form pacts that are mutually beneficial.

Helping closely related organisms can appear altruistic; but this is also a selfish behavior. Reproductive success (fitness) has two components; direct fitness and indirect fitness. Direct fitness is a measure of how many alleles, on average, a genotype contributes to the subsequent generation's gene pool by reproducing. Indirect fitness is a measure of how many alleles identical to its own it helps to enter the gene pool. Direct fitness plus indirect fitness is inclusive fitness. J. B. S. Haldane once remarked he would gladly drown, if by doing so he saved two siblings or eight cousins. Each of his siblings would share one half his alleles; his cousins, one eighth. They could potentially add as many of his alleles to the gene pool as he could.

Natural selection favors traits or behaviors that increase a genotype's inclusive fitness. Closely related organisms share many of the same alleles. In diploid species, siblings share on average at least 50% of their alleles. The percentage is higher if the parents are related. So, helping close relatives to reproduce gets an organism's own alleles better represented in the gene pool. The benefit of helping relatives increases dramatically in highly inbred species. In some cases, organisms will completely forgo reproducing and only help their relatives reproduce. Ants, and other eusocial insects, have sterile castes that only serve the queen and assist her reproductive efforts. The sterile workers are reproducing by proxy.

The words selfish and altruistic have connotations in everyday use that biologists do not intend. Selfish simply means behaving in such a way that one's own inclusive fitness is maximised; altruistic means behaving in such a way that another's fitness is increased at the expense of ones' own. Use of the words selfish and altruistic is not meant to imply that organisms consciously understand their motives. The opportunity for natural selection to operate does not induce genetic variation to appear — selection only distinguishes between existing variants. Variation is not possible along every imaginable axis, so all possible adaptive solutions are not open to populations. To pick a somewhat ridiculous example, a steel shelled turtle might be an improvement over regular turtles. Turtles are killed quite a bit by cars these days because when confronted with danger, they retreat into their shells — this is not a great strategy against a two ton automobile. However, there is no variation in metal content of shells, so it would not be possible to select for a steel shelled turtle.

Here is a second example of natural selection. Geospiza fortis lives on the Galapagos islands along with fourteen other finch species. It feeds on the seeds of the plant Tribulus cistoides, specialising on the smaller seeds. Another species, G. Magnirostris, has a larger beak and specialises on the larger seeds. The health of these bird populations depends on seed production. Seed production, in turn, depends on the arrival of wet season. In 1977, there was a drought. Rainfall was well below normal and fewer seeds were produced. As the season progressed, the G. fortis population depleted the supply of small seeds. Eventually, only larger seeds remained.

Most of the finches starved; the population plummeted from about twelve hundred birds to less than two hundred. Peter Grant, who had been studying these finches, noted that larger beaked birds fared better than smaller beaked ones. These larger birds had offspring with correspondingly large beaks. Thus, there was an increase in the proportion of large beaked birds in the population the next generation. To prove that the change in bill size in Geospiza fortis was an evolutionary change, Grant had to show that differences in bill size were at least partially genetically based. He did so by crossing finches of various beak sizes and showing that a finch's beak size was influenced by its parent's genes. Large beaked birds had large beaked offspring; beak size was not due to environmental differences.

Natural selection may not lead a population to have the optimal set of traits. In any population, there would be a certain combination of possible alleles that would produce the optimal set of traits (the global optimum); but there are other sets of alleles that would yield a population almost as adapted (local optima). Transition from a local optimum to the global optimum may be hindered or forbidden because the population would have to pass through less adaptive states to make the transition. Natural selection only works to bring populations to the nearest optimal point. This idea is Sewall Wright's adaptive landscape. This is one of the most influential models that shape how evolutionary biologists view evolution.

Natural selection does not have any foresight. It only allows organisms to adapt to their current environment. Structures or behaviors do not evolve for future utility. An organism adapts to its environment at each stage of its evolution. As the environment changes, new traits may be selected for. Large changes in populations are the result of cumulative natural selection. Changes are introduced into the population by mutation; the small minority of these changes that result in a greater reproductive output of their bearers are amplified in frequency by selection.

Complex traits must evolve through viable intermediates. For many traits, it initially seems unlikely that intermediates would be viable. What good is half a wing? Half a wing may be no good for flying, but it may be useful in other ways. Feathers are thought to have evolved as insulation (ever worn a down jacket?) and/or as a way to trap insects. Later, proto-birds may have learned to glide when leaping from tree to tree. Eventually, the feathers that originally served as insulation now became co-opted for use in flight. A trait's current utility is not always indicative of its past utility. It can evolve for one purpose, and be used later for another. A trait evolved for its current utility is an adaptation; one that evolved for another utility is an exaptation. An example of an exaptation is a penguin's wing. Penguins evolved from flying ancestors; now they are flightless and use their wings for swimming.

Sexual Selection

In many species, males develop prominent secondary sexual characteristics. A few oft cited examples are the peacock's tail, coloring and patterns in male birds in general, voice calls in frogs and flashes in fireflies. Many of these traits are a liability from the standpoint of survival. Any ostentatious trait or noisy, attention getting behavior will alert predators as well as potential mates. Sexual selection is natural selection operating on factors that contribute to an organism's mating success. Traits that are a liability to survival can evolve when the sexual attractiveness of a trait outweighs the liability incurred for survival. A male who lives a short time, but produces many offspring is much more successful than a long lived one that produces few. The former's genes will eventually dominate the gene pool of his species. In many species, especially polygynous species where only a few males monopolise all the females, sexual selection has caused pronounced sexual dimorphism. In these species males compete against other males for mates. The competition can be either direct or mediated by female choice. In species where females choose, males compete by displaying striking phenotypic characteristics and/or performing elaborate courtship behaviors. The females then mate with the males that most interest them, usually the ones with the most outlandish displays. There are many competing theories as to why females are attracted to these displays.

The good genes model states that the display indicates some component of male fitness. A good genes advocate would say that bright coloring in male birds indicates a lack of parasites. The females are cueing on some signal that is correlated with some other component of viability. Selection for good genes can be seen in sticklebacks. In these fish, males have red coloration on their sides. Milinski and Bakker showed that intensity of color was correlated to both parasite load and sexual attractiveness. Females preferred redder males. The redness indicated that he was carrying fewer parasites.

Evolution can get stuck in a positive feedback loop. Another model to explain secondary sexual characteristics is called the runaway sexual selection model. R. A. Fisher proposed that females may have an innate preference for some male trait before it appears in a population. Females would then mate with male carriers when the trait appears. The offspring of these matings have the genes for both the trait and the preference for the trait. As a result, the process snowballs until natural selection brings it into check. Suppose that female birds prefer males with longer than average tail feathers. Mutant males with longer than average feathers will produce more offspring than the short feathered males.

In the next generation, average tail length will increase. As the generations progress, feather length will increase because females do not prefer a specific length tail, but a longer than average tail. Eventually tail length will increase to the point were the liability to survival is matched by the sexual attractiveness of the trait and an equilibrium will be established. Note that in many exotic birds male plumage is often very showy and many species do in fact have males with greatly elongated feathers. In some cases these feathers are shed after the breeding season. None of the above models are mutually exclusive. There are millions of sexually dimorphic species on this planet and the forms of sexual selection probably vary amongst them.

Genetic Drift

Allele frequencies can change due to chance alone. This is called genetic drift. Drift is a binomial sampling error of the gene pool. What this means is, the alleles that form the next generation's gene pool are a sample of the alleles from the current generation. When sampled from a population, the frequency of alleles differs slightly due to chance alone. Alleles can increase or decrease in frequency due to drift. The average expected change in allele frequency is zero, since increasing or decreasing in frequency is equally probable. A small percentage of alleles may continually change frequency in a single direction for several generations just as flipping a fair coin may, on occasion, result in a string of heads or tails. A very few new mutant alleles can drift to fixation in this manner. In small populations, the variance in the rate of change of allele frequencies is greater than in large populations. However, the overall rate of genetic drift (measured in substitutions per generation) is independent of population size. If the mutation rate is constant, large and small populations lose alleles to drift at the same rate. This is because large populations will have more alleles in the gene pool, but they will lose them more slowly. Smaller populations will have fewer alleles, but these will quickly cycle through. This assumes that mutation is constantly adding new alleles to the gene pool and selection is not operating on any of these alleles.

Sharp drops in population size can change allele frequencies substantially. When a population crashes, the alleles in the surviving sample may not be representative of the precrash gene pool. This change in the gene pool is called the founder effect, because small populations of organisms that invade a new territory (founders) are subject to this. Many biologists feel the genetic changes brought about by founder effects may contribute to isolated populations developing reproductive isolation from their parent populations. In sufficiently small populations, genetic drift can counteract selection. Mildly deleterious alleles may drift to fixation.

Wright and Fisher disagreed on the importance of drift. Fisher thought populations were sufficiently large that drift could be neglected. Wright argued that populations were often divided into smaller subpopulations. Drift could cause allele frequency differences between subpopulations if gene flow was small enough. If a subpopulation was small enough, the population could even drift through fitness valleys in the adaptive landscape. Then, the subpopulation could climb a larger fitness hill. Gene flow out of this subpopulation could contribute to the population as a whole adapting. This is Wright's Shifting Balance theory of evolution. Both natural selection and genetic drift decrease genetic variation. If they were the only mechanisms of evolution, populations would eventually become homogeneous and further evolution would be impossible. There are, however, mechanisms that replace variation depleted by selection and drift.

Increasing Genetic Variation Mechanisms

Mutation

The cellular machinery that copies DNA sometimes makes mistakes. These mistakes alter the sequence of a gene. This is called a mutation. There are many kinds of mutations. A point mutation is a mutation in which one "letter" of the genetic code is changed to another. Lengths of DNA can also be deleted or inserted in a gene; these are also mutations. Finally, genes or parts of genes can become inverted or duplicated. Typical rates of mutation are between 10-10 and 10-12 mutations per base pair of DNA per generation. Most mutations are thought to be neutral with regards to fitness. (Kimura defines neutral as $|s| < 1/2Ne$, where s is the selective coefficient and Ne is the effective population size.) Only a small portion of the genome of eukaryotes contains coding segments. And, although some non-coding DNA is involved in gene regulation or other cellular functions, it is probable that most base changes would have no fitness consequence. Most mutations that have any phenotypic effect are deleterious. Mutations that result in amino acid substitutions can change the shape of a protein, potentially changing or eliminating its function. This can lead to inadequacies in biochemical pathways or interfere with the process of development. Organisms are sufficiently integrated that most random changes will not produce a fitness benefit. Only a very small percentage of mutations are beneficial. The ratio of neutral to deleterious to beneficial mutations is unknown and probably varies with respect to details of the locus in question and environment.

Mutation limits the rate of evolution. The rate of evolution can be expressed in terms of nucleotide substitutions in a lineage per generation. Substitution is the replacement of an allele by another in a population. This is a two step process: First a mutation occurs in an individual, creating a new allele. This allele subsequently increases in frequency to fixation in the population. The rate of evolution is $k = 2Nvu$ (in diploids) where k is nucleotide substitutions, N is the effective population size, v is the rate of mutation and u is the proportion of mutants that eventually fix in the population. Mutation need not be limiting over short time spans. The rate of evolution expressed above is given as a steady state equation; it assumes the system is at equilibrium. Given the time frames for a single mutant to fix, it is unclear if populations are ever at equilibrium. A change in environment can cause previously neutral alleles to have selective values; in the short term evolution can run on "stored" variation and thus is independent of mutation rate. Other mechanisms can also contribute selectable variation. Recombination creates new combinations of alleles (or new alleles) by joining sequences with separate microevolutionary histories within a population.

Mutant Alleles

Mutation creates new alleles. Each new allele enters the gene pool as a single copy amongst many. Most are lost from the gene pool, the organism carrying them fails to reproduce, or reproduces but does not pass on that particular allele. A mutant's fate is shared with the genetic background it appears in. A new allele will initially be linked to other loci in its genetic background, even loci on other chromosomes. If the allele increases in frequency in the population, initially it will be paired with other alleles at that locus — the new allele will primarily be carried in individuals heterozygous for that locus. The chance of it being paired with itself is low until it reaches intermediate frequency. If the allele is recessive, its effect won't be seen in any individual until a homozygote is formed. The eventual fate of the allele depends on whether it is neutral, deleterious or beneficial.

Neutral alleles

Most neutral alleles are lost soon after they appear. The average time (in generations) until loss of a neutral allele is 2(Ne/N) ln(2N) where N is the effective population size (the number of individuals contributing to the next generation's gene pool) and N is the total population size. Only a small percentage of alleles fix. Fixation is the process of an allele increasing to a frequency at or near one. The probability of a neutral allele fixing in a population is equal to its frequency. For a new mutant in a diploid population, this frequency is 1/2N. If mutations are neutral with respect to fitness, the rate of substitution (k) is equal to the rate of mutation(v). This does not mean every new mutant eventually reaches fixation. Alleles are added to the gene pool by mutation at the same rate they are lost to drift. For neutral alleles that do fix, it takes an average of 4N generations to do so. However, at equilibrium there are multiple alleles segregating in the population. In small populations, few mutations appear each generation. The ones that fix do so quickly relative to large populations. In large populations, more mutants appear over the generations. But, the ones that fix take much longer to do so. Thus, the rate of neutral evolution (in substitutions per generation) is independent of population size.

The rate of mutation determines the level of heterozygosity at a locus according to the neutral theory. Heterozygosity is simply the proportion of the population that is heterozygous. Equilibrium heterozygosity is given as H = 4Nv/[4Nv+1] (for diploid populations). H can vary from a very small number to almost one. In small populations, H is small (because the equation is approximately a very small number divided by one). In (biologically unrealistically) large populations, heterozygosity approaches one (because the equation is approximately a large number divided by itself). Directly testing this model is difficult because N and v can only be estimated for most natural populations. But, heterozygosities are believed to be too low to be described by a strictly neutral model. Solutions offered by neutralists for this discrepancy include hypothesizing that natural populations may not be at equilibrium.

At equilibrium there should be a few alleles at intermediate frequency and many at very low frequencies. This is the Ewens- Watterson distribution. New alleles enter a population every generation, most remain at low frequency until they are lost. A few drift to intermediate frequencies, a very few drift all the way to fixation. In *Drosophila pseudoobscura*, the protein Xanthine dehydrogenase (Xdh) has many variants. In a single population, Keith, et. al., found that 59 of 96 proteins were of one type, two others were represented ten and nine times and nine other types were present singly or in low numbers.

Deleterious alleles

Deleterious mutants are selected against but remain at low frequency in the gene pool. In diploids, a deleterious recessive mutant may increase in frequency due to drift.

Selection cannot see it when it is masked by a dominant allele. Many disease causing alleles remain at low frequency for this reason. People who are carriers do not suffer the negative effect of the allele. Unless they mate with another carrier, the allele may simply continue to be passed on. Deleterious alleles also remain in populations at a low frequency due to a balance between recurrent mutation and selection. This is called the mutation load.

Beneficial alleles

Most new mutants are lost, even beneficial ones. Wright calculated that the probability of fixation of a beneficial allele is 2s. (This assumes a large population size, a small fitness benefit, and that heterozygotes have an intermediate fitness. A benefit of 2s yields an overall rate of evolution: k=4Nvs where v is the mutation rate to beneficial alleles) An allele that conferred a one percent increase in fitness only has a two percent chance of fixing. The probability of fixation of beneficial type of mutant is boosted by recurrent mutation. The beneficial mutant may be lost several times, but eventually it will arise and stick in a population. (Recall that even deleterious mutants recur in a population.)

Directional selection depletes genetic variation at the selected locus as the fitter allele sweeps to fixation. Sequences linked to the selected allele also increase in frequency due to hitchhiking. The lower the rate of recombination, the larger the window of sequence that hitchhikes. Begun and Aquadro compared the level of nucleotide polymorphism within and between species with the rate of recombination at a locus. Low levels of nucleotide polymorphism within species coincided with low rates of recombination. This could be explained by molecular mechanisms if recombination itself was mutagenic. In this case, recombination with also be correlated with nucleotide divergence between species. But, the level of sequence divergence did not correlate with the rate of recombination. Thus, they inferred that selection was the cause. The correlation between recombination and nucleotide polymorphism leaves the conclusion that selective sweeps occur often enough to leave an imprint on the level of genetic variation in natural populations.

Recombination

Each chromosome in our sperm or egg cells is a mixture of genes from our mother and our father. Recombination can be thought of as gene shuffling. Most organisms have linear chromosomes and their genes lie at specific location (loci) along them. Bacteria have circular chromosomes. In most sexually reproducing organisms, there are two of each chromosome type in every cell. For instance in humans, every chromosome is paired, one inherited from the mother, the other inherited from the father. When an organism produces gametes, the gametes end up with only one of each chromosome per cell. Haploid gametes are produced from diploid cells by a process called meiosis. In meiosis, homologous chromosomes line up. The DNA of the chromosome is broken on both chromosomes in several places and rejoined with the other strand. Later, the two homologous chromosomes are split into two separate cells that divide and become gametes. But, because of recombination, both of the chromosomes are a mix of alleles from the mother and father. Recombination creates new combinations of alleles. Alleles that arose at different times and different places can be brought together. Recombination can occur not only between genes, but within genes as well. Recombination within a gene can form a new allele. Recombination is a mechanism of evolution because it adds new alleles and combinations of alleles to the gene pool.

Gene Flow

New organisms may enter a population by migration from another population. If they mate within the population, they can bring new alleles to the local gene pool. This is called gene flow. In some closely related species, fertile hybrids can result from interspecific matings. These hybrids can vector genes from species to species.

Gene flow between more distantly related species occurs infrequently. This is called horizontal transfer. One interesting case of this involves genetic elements called P elements. Margaret Kidwell found that P elements were transferred from some species in the *Drosophila willistoni* group to *Drosophila melanogaster*. These two species of fruit flies are distantly related and hybrids do not form. Their ranges do, however, overlap. The P elements were vectored into *D. melanogaster* via a parasitic mite that targets both these species. This mite punctures the exoskeleton of the flies and feeds on the "juices". Material, including DNA, from one fly can be transferred to another when the mite feeds. Since P elements actively move in the genome (they are themselves parasites of DNA), one incorporated itself into the genome of a *melanogaster* fly and subsequently spread through the species. Laboratory stocks of *melanogaster* caught prior to the 1940's lack of P elements. All natural populations today harbor them.

Theory of Evolution

Biology came of age as a science when Charles Darwin published "On the Origin of Species." But, the idea of evolution wasn't new to Darwin. Lamarck published a theory of evolution in 1809. Lamarck thought that species arose continually from nonliving sources. These species were initially very primitive, but increased in complexity over time due to some inherent tendency. This type of evolution is called orthogenesis. Lamarck proposed that an organism's acclimation to the environment could be passed on to its offspring. For example, he thought proto-giraffes stretched their necks to reach higher twigs. This caused their offspring to be born with longer necks. This proposed mechanism of evolution is called the inheritance of acquired characteristics. Lamarck also believed species never went extinct, although they may change into newer forms. All three of these ideas are now known to be wrong.

Darwin's contributions include hypothesising the pattern of common descent and proposing a mechanism for evolution — natural selection. In Darwin's theory of natural selection, new variants arise continually within populations. A small percentage of these variants cause their bearers to produce more offspring than others. These variants thrive and supplant their less productive competitors. The effect of numerous instances of selection would lead to a species being modified over time. Darwin's theory did not accord with older theories of genetics. In Darwin's time, biologists held to the theory of blending inheritance — an offspring was an average of its parents. If an individual had one short parent and one tall parent, it would be of medium height. And, the offspring would pass on genes for medium sized offspring. If this was the case, new genetic variations would quickly be diluted out of a population. They could not accumulate as the theory of evolution required. We now know that the idea of blending inheritance is wrong.

Darwin didn't know that the true mode of inheritance was discovered in his lifetime. Gregor Mendel, in his experiments on hybrid peas, showed that genes from a mother and father do not blend. An offspring from a short and a tall parent may be medium sized; but it carries genes for shortness and tallness. The genes remain distinct and can be passed on to subsequent generations. Mendel mailed his paper to Darwin, but Darwin never opened it. It was a long time until Mendel's ideas were accepted. One group of biologists, called biometricians, thought Mendel's laws only held for a few traits. Most traits, they claimed, were governed by blending inheritance. Mendel studied discrete traits. Two of the traits in his famous experiments were smooth versus wrinkled coat on peas. This trait did not vary continuously. In other words, peas are either wrinkled or smooth — intermediates are not found. Biometricians considered these traits aberrations. They studied continuously varying traits like size and believed most traits showed blending inheritance.

The discrete genes Mendel discovered would exist at some frequency in natural populations. Biologists wondered how and if these frequencies would change. Many thought that the more common versions of genes would increase in frequency simply because they were already at high frequency. Hardy and Weinberg independently showed that the frequency of an allele would not change over time simply due to its being rare or common. Their model had several assumptions — that all alleles reproduced at the same rate, that the population size was very large and that alleles did not change in form. Later, R. A. Fisher showed that Mendel's laws could explain continuous traits if the expression of these traits were due to the action of many genes. After this, geneticists accepted Mendel's Laws as the basic rules of genetics. From this basis, Fisher, Sewall Wright and J. B. S.. Haldane founded the field of population genetics. Population genetics is a field of biology that attempts to measure and explain the levels of genetic variation in populations. R. A. Fisher studied the effect of natural selection on large populations. He demonstrated that even very small selective differences amongst alleles could cause appreciable changes in allele frequencies over time. He also showed that the rate of adaptive change in a population is proportional to the amount of genetic variation present. This is called Fisher's Fundamental Theorem of Natural Selection. Although it is called the fundamental theorem, it does not hold in all cases. The rate at which natural selection brings about adaptation depends on the details of how selection is working. In some rare cases, natural selection can actually cause a decline in the mean relative fitness of a population.

Sewall Wright was more concerned with drift. He stressed that large populations are often subdivided into many subpopulations. In his theory, genetic drift played a more important role compared to selection. Differentiation between subpopulations, followed by migration among them, could contribute to adaptations amongst populations. Wright also came up with the idea of the adaptive landscape — an idea that remains influential to this day. Its influence remains even though P. A. P. Moran has shown that, mathematically, adaptive landscapes don't exist as Wright envisioned them. Wright extended his results of one-locus models to a two-locus case in proposing the adaptive landscape. But, unbeknownst to him, the general conclusions of the one-locus model don't extend to the two-locus case.

J. B. S. Haldane developed many of the mathematical models of natural and artificial selection. He showed that selection and mutation could oppose each other, that deleterious mutations could remain in a population due to recurrent mutation. He also demonstrated that there was a cost to natural selection, placing a limit on the amount of adaptive substitutions a population could undergo in a given time frame. For a long time, population genetics developed as a theoretical field. But, gathering the data needed to test the theories was nearly impossible. Prior to the advent of molecular biology, estimates of genetic variability could only be inferred from levels of morphological differences in populations. Lewontin and Hubby were the first to get a good estimate of genetic variation in a population. Using the then new technique of protein electrophoresis, they showed that 30% of the loci in a population of *Drosophila pseudoobscura* were polymorphic. They also showed that it was likely that they could not detect all the variation that was present.

Motoo Kimura theorised that most variation found in populations was selectively equivalent (neutral). Multiple alleles at a locus differed in sequence, but their fitnesses were the same. Kimura's neutral theory described rates of evolution and levels of polymorphism solely in terms of mutation and genetic drift. The neutral theory did not deny that natural selection acted on natural populations; but it claimed that the majority of natural variation was transient polymorphisms of neutral alleles. Selection did not act frequently or strongly enough to influence rates of evolution or levels of polymorphism. Initially, a wide variety of observations seemed to be consistent with the neutral theory. Eventually, however, several lines of evidence toppled it. There is less variation in natural populations than the neutral theory predicts. Also, there is too much variance in rates of substitutions in different lineages to be explained by mutation and drift alone. Finally, selection itself has been shown to have an impact on levels of nucleotide variation. Currently, there is no comprehensive mathematical theory of evolution that accurately predicts rates of evolution and levels of heterozygosity in natural populations.

Pattern of Macroevolution

Evolution is not progress. The popular notion that evolution can be represented as a series of improvements from simple cells, through more complex life forms, to humans (the pinnacle of evolution), can be traced to the concept of the scale of nature.All species have descended from a common ancestor. As time went on, different lineages of organisms were modified with descent to adapt to their environments. Thus, evolution is best viewed as a branching tree or bush, with the tips of each branch representing currently living species. No living organisms today are our ancestors. Every living species is as fully modern as we are with its own unique evolutionary history. No extant species are "lower life forms," atavistic stepping stones paving the road to humanity. A related, and common, fallacy about evolution is that humans evolved from some living species of ape. This is not the case — humans and apes share a common ancestor. Both humans and living apes are fully modern species; the ancestor we evolved from was an ape, but it is now extinct and was not the same as present day apes (or humans for that matter). If it were not for the vanity of human beings, we would be classified as an ape. Our closest relatives are, collectively, the chimpanzee and the pygmy chimp. Our next nearest relative is the gorilla.

Macroevolution is studied by examining patterns in biological populations and groups of related organisms and inferring process from pattern. Given the observation of microevolution and the knowledge that the earth is billions of years old — macroevolution could be postulated. But this extrapolation, in and of itself, does not provide a compelling explanation of the patterns of biological diversity we see today. Evidence for macroevolution, or common ancestry and modification with descent, comes from several other fields of study. These include: comparative biochemical and genetic studies, comparative developmental biology, patterns of biogeography, comparative morphology and anatomy and the fossil record. Closely related species (as determined by morphologists) have similar gene sequences. Overall sequence similarity is not the whole story, however. The pattern of differences we see in closely related genomes is worth examining. All living organisms use DNA as their genetic material, although some viruses use RNA. DNA is composed of strings of nucleotides. There are four different kinds of nucleotides: adenine (A), guanine (G), cytosine (C) and thymine (T). Genes are sequences of nucleotides that code for proteins. Within a gene, each block of three nucleotides is called a codon. Each codon designates an amino acid (the subunits of proteins). The three letter code is the same for all organisms (with a few exceptions). There are 64 codons, but only 20 amino acids to code for; so, most amino acids are coded for by several codons. In many cases the first two nucleotides in the codon designate the amino acid. The third position can have any of the four nucleotides and not effect how the code is translated. A gene, when in use, is transcribed into RNA — a nucleic acid similar to DNA. (RNA, like DNA, is made up of nucleotides although t he nucleotide uracil (U) is used in place of thymine (T).) The RNA transcribed from a gene is called messenger RNA. Messenger RNA is then translated via cellular machinery called ribosomes into a string of amino acids — a protein. Some proteins function as enzymes, catalysts that speed the chemical reactions in cells. Others are structural or involved in regulating development.

Gene sequences in closely related species are very similar. Often, the same codon specifies a given amino acid in two related species, even though alternate codons could serve functionally as well. But, some differences do exist in gene sequences. Most often, differences are in third codon positions, where changes in the DNA sequence would not disrupt the sequence of the protein. There are other sites in the genome where nucleotide differences do not effect protein sequences. The genome of eukaryotes is loaded with 'dead genes' called pseudogenes. Pseudogenes are copies of working genes that have been inactivated by mutation. Most pseudogenes do not produce full proteins. They may be transcribed, but not translated. Or, they may be translated, but only a truncated protein is produced. Pseudogenes evolve much faster than their working counterparts. Mutations in them do not get incorporated into proteins, so they have no effect on the fitness of an organism.

Introns are sequences of DNA that interrupt a gene, but do not code for anything. The coding portions of a gene are called exons. Introns are spliced out of the messenger RNA prior to translation, so they do not contribute information needed to make the protein. They are sometimes, however, involved in regulation of the gene. Like pseudogenes, introns (in general) evolve faster than coding portions of a gene. Nucleotide positions that can be changed without changing the sequence of a protein are called silent sites. Sites where changes result in an amino acid substitution are called replacement sites. Silent sites are expected to be more polymorphic within a population and show more differences between populations. Although both silent and replacement sites receive the same amount of mutations, natural selection only infrequently allows changes at replacement sites. Silent sites, however, are not as constrained.

Kreitman was the first demonstrate that silent sites were more variable than coding sites. Shortly after the methods of DNA sequencing were discovered, he sequenced 11 alleles of the enzyme alcohol dehydrogenase (AdH). Of the 43 polymorphic nucleotide sites he found, only one resulted in a change in the amino acid sequence of the protein. Silent sites may not be entirely selectively neutral. Some DNA sequences are involved with regulation of genes, changes in these sites may be deleterious. Likewise, although several codons code for a single amino acid, an organism may have a preferred codon for each amino acid. This is called codon bias. If two species shared a recent common ancestor one would expect genetic information, even information such as redundant nucleotides and the position of introns or pseudogenes, to be similar. Both species would have inherited this information from their common ancestor.

The degree of similarity in nucleotide sequence is a function of divergence time. If two populations had recently separated, few differences would have built up between them. If they separated long ago, each population would have evolved numerous differences from their common ancestor (and each other). The degree of similarity would also be a function of silent versus replacement sites. Li and Graur, in their molecular evolution text, give the rates of evolution for silent vs. replacement rates. The rates were estimated from sequence comparisons of 30 genes from humans and rodents, which diverged about 80 million years ago. Silent sites evolved at an average rate of 4.61 nucleotide substitution per 109 years. Replacement sites evolved much slower at an average rate of 0.85 nucleotide substitutions per 109 years.

Groups of related organisms are 'variations on a theme' — the same set of bones are used to construct all vertebrates. The bones of the human hand grow out of the same tissue as the bones of a bat's wing or a whale's flipper; and, they share many identifying features such as muscle insertion points and ridges. The only difference is that they are scaled differently. Evolutionary biologists say this indicates that all mammals are modified descendants of a common ancestor which had the same set of bones.

Closely related organisms share similar developmental pathways. The differences in development are most evident at the end. As organisms evolve, their developmental pathway gets modified. An alteration near the end of a developmental pathway is less likely to be deleterious than changes in early development. Changes early on may have a cascading effect. Thus most evolutionary changes in development are expected to take place at the periphery of development, or in early aspects of development that have no later repercussions. For a change in early development to be propagated, the benefit of the early alteration must outweigh the consequences to later development.

Because they have evolved this way, organisms pass through the early stages of development that their ancestors passed through up to the point of divergence. So, an organism's development mimics its ancestors although it doesn't recreate it exactly. Development of the flatfish, Pleuronectes, illustrates this point. Early on, Pleuronectes develops a tail that comes to a point. In the next developmental stage, the top lobe of the tail is larger than the bottom lobe (as in sharks). When development is complete, the upper and lower lobes are equally sized. This developmental pattern mirrors the evolutionary transitions it has undergone. Natural selection can modify any stage of a life cycle, so some differences are seen in early development. Thus, evolution does not always recapitulate ancestral forms — butterflies did not evolve from ancestral caterpillars, for example. There are differences in the appearance of early vertebrate embryos. Amphibians rapidly form a ball of cells in early development. Birds, reptiles and mammals form a disk. The shape of the early embryo is a result of different yolk concentrations in the eggs. Birds' and reptiles' eggs are heavily yolked. Their eggs develop similarly to amphibians except the yolk has deformed the shape of the embryo. The ball is stretched out and lying atop the yolk. Mammals have no yolk, but still form a disk early. This is because they have descended from reptiles. Mammals lost their yolky eggs, but retained the early pattern of development. In all these vertebrates, the pattern of cell movements is similar despite superficial differences in appearance. In addition, all types quickly converge upon a primitive, fish-like stage within a few days. From there, development diverges.

Traces of an organism's ancestry sometimes remain even when an organism's development is complete. These are called vestigial structures. Many snakes have rudimentary pelvic bones retained from their walking ancestors. Vestigial does not mean useless, it means the structure is clearly a vestige of an structure inherited from ancestral organisms. Vestigial structures may acquire new functions. In humans, the appendix now houses some immune system cells. Closely related organisms are usually found in close geographic proximity; this is especially true of organisms with limited dispersal opportunities. The mammalian fauna of Australia is often cited as an example of this; marsupial mammals fill most of the equivalent niches that placentals fill in other ecosystems. If all organisms descended from a common ancestor, species distribution across the planet would be a function of site of origination, potential for dispersal, distribution of suitable habitat, and time since origination. In the case of Australian mammals, their physical separation from sources of placentals means potential niches were filled by a marsupial radiation rather than a placental radiation or invasion.

Natural selection can only mold available genetically based variation. In addition, natural selection provides no mechanism for advance planning. If selection can only tinker with the available genetic variation, we should expect to see examples of jury-rigged design in living species. This is indeed the case. In lizards of the genus *Cnemodophorus*, females reproduce parthenogenetically. Fertility in these lizards is increased when a female mounts another female and simulates copulation. These lizards evolved from sexual lizards whose hormones were aroused by sexual behavior. Now, although the sexual mode of reproduction has been lost, the means of getting aroused (and hence fertile) has been retained. Fossils show hard structures of organisms less and less similar to modern organisms in progressively older rocks. In addition, patterns of biogeography apply to fossils as well as extant organisms. When combined with plate tectonics, fossils provide evidence of distributions and dispersals of ancient species. For example, South America had a very distinct marsupial mammalian fauna until the land bridge formed between North and South America. After that marsupials started disappearing and placentals took their place. This is commonly interpreted as the placentals wiping out the marsupials, but this may be an over simplification.

Transitional fossils between groups have been found. One of the most impressive transitional series is the ancient reptile to modern mammal transition. Mammals and reptiles differ in skeletal details, especially in their skulls. Reptilian jaws have four bones. The foremost is called the dentary. In mammals, the dentary bone is the only bone in the lower jaw. The other bones are part of the middle ear. Reptiles have a weak jaw and a mouthful of undifferentiated teeth. Their jaw is closed by three muscles: the external, posterior and internal adductor. Each reptile tooth is single cusped. Mammals have powerful jaws with differentiated teeth. Many of these teeth, such as the molars, are multi-cusped. The temporalis and masseter muscles, derived from the external adductor, close the mammalian jaw. Mammals have a secondary palate, a bony structure separating their nostril passages and throat, so most can swallow and breathe simultaneously. Reptiles lack this.

The evolution of these traits can be seen in a series of fossils. *Procynosuchus* shows an increase in size of the dentary bone and the beginnings of a palate. *Thrinaxodon* has a reduced number of incisors, a precursor to tooth differentiation. *Cynognathus* (a doglike carnivore) shows a further increase in size of the dentary bone. The other three bones are located inside the back portion of the jaw. Some teeth are multicusped and the teeth fit together tightly. *Diademodon* (a plant eater) shows a more advanced degree of occlusion (teeth fitting tightly). *Probelesodon* has developed a double joint in the jaw. The jaw could hinge off two points with the upper skull. The front hinge was probably the actual hinge while the rear hinge was an alignment guide.

The forward movement of a hinge point allowed for the precursor to the modern masseter muscle to anchor further forward in the jaw. This allowed for a more powerful bite. The first true mammal was *Morgonucudon,* a rodent-like insectivore from the late Triassic. It had all the traits common to modern mammals. These species were not from a single, unbranched lineage. Each is an example from a group of organisms along the main line of mammalian ancestry. The strongest evidence for macroevolution comes from the fact that suites of traits in biological entities fall into a nested pattern. For example, plants can be divided into two broad categories, non- vascular (ex. mosses) and vascular. Vascular plants can be divided into seedless (ex. ferns) and seeded. Vascular seeded plants can be divided into gymnosperms (ex. pines) and flowering plants (angiosperms). Angiosperms can be divided into monocots and dicots. Each of these types of plants have several characters that distinguish them from other plants. Traits are not mixed and matched in groups of organisms.

EVOLUTIONAL AND MOLECULAR BIOLOGY

All present day life is based on a nucleic acid informational molecule, either DNA or RNA that encodes information needed to make a living cell. The information is coded in a triplet code and though there are some examples of slight variations in this code, no radical departures from it exist. This "universal" code is interpreted as proteins by a complex machinery called the ribosome that is also shared in common among all living things. These main features of information storage and retrieval are conserved and provide convincing evidence that all life on earth has one origin and shares a common ancestor. Since life began, it has been changing in permissible directions. Physical constraints on the chemistry of life including the properties of water, the nature of carbon and other key aspects of biology have allowed variations on the original theme, but only within certain boundaries. However, 3.85 billion years is a long time and many variations have been tried out and many have been successful. Because the informational molecule has to transmit the information through time for the blueprint of a cell, it is the information that has changed. This molecule has been copied billions of times, but not without some errors creeping in.

Today, we can look back in time by comparing the sequences of the nucleotides and the translated sequences of the proteins from present day organisms. By quantitating the differences between the sequences and by making some modest assumptions about rates of change in the sequences, we can estimate when different organisms diverged from one another. These dates can be calibrated by the fossil record which can be dated to high precision. Very similar organisms have very similar sequences, and more distant relatives have more differences in their code. This is the concept of the molecular clock. With this very simple idea, one can use sequences to build trees with branches representing different species.

The relationships between organisms can be graphed in this way. If enough organisms are included and the most conserved sequences are used, a tree of life can be constructed. This is an approximate genealogy of present day organisms, and it is very interesting. In your handout look at the tree for the eukaryotes This was made from joining four different proteins together and making a longer super protein for sequence analysis. The proteins used are EF1 alpha, alpha and beta tubulin and actin. These are highly conserved eukaryotic proteins suitable for this tree making purpose. In such a tree, the branches always diverge, they do not merge back together, because species do not merge, except in very rare events. The place where two branches come together is the point in time when they were the same species. Farther and farther back on the tree are divergences that are deeper and more ancient. If we go back far enough, the most distant branches meet, at the common ancestor.

The single celled organism that gave rise to all life on the planet. This model of evolution has been challenged lately by evidence that early bacteria shared their DNA often, leading to a network topology rather than a branching bush. This sharing of genetic material has become less common as the different lineages have become less similar to each other. However, this makes any effort to go back and identify the common ancestor impossible, because there were fuzzy definitions of species at these early times. To build a tree like this is not trivial. It takes some care to pick the right sequences to use, because not all sequences are appropriate for this job. Even in a single gene, not all of the gene is useful for this task. Frequently, only the most conserved parts of a gene are included in building trees. Ideally, the most ancient features common to all life would be good candidates to compare. This has been done most often with ribosomal RNA, since all life has this molecule and it is so central to life that it has to be highly conserved. This is the basis of the rRNA database project. Here 57,773 rRNA sequences have been used to make trees. These trees include every kind of life possible to get a pretty comprehensive tree of life. The results look like the tree in your handout. Here representatives from the major groups of organisms are shown. Every other organism will fit on this tree very close to an existing branch, so it does little good to use all the sequences, because it just clutters up the picture. It is clear from this tree that there are three main divisions in life. These have been called domains.

In a hierarchy of life, domains are higher than kingdoms. The three domains are bacteria, archaea and eukarya. The taxonomy of life at these highest levels is a matter of philosophy. Ernst Mayr, one of the most prominent of all evolutionary biologists, does not accept the three domain system. He argues that the fundamental division in life is the prokaryote eukaryote division and that the archaea are just another group of bacteria that do not deserve domain status. Bacteria and archaea are both prokaryotes, without a nucleus. They are as different from each other as each is to the eukarya, so even though they are prokaryotes, it is difficult to lump them together. In fact, the most common version of this tree shows the archaea to be more closely related to eukarya, but this is a debated issue. Researchers into the early evolution of life are actively trying to resolve this issue, but it is proving to be very hard. The evidence accumulating now shows that archea are a mix of bacterial and eukaryotic like genes.

The last common ancestor was a prokaryote, with a cell wall and a lipid bilayer, incorporating membrane proteins within it. This organism probably used electron transport and proton pumping to make a proton gradient for ATP synthesis. This strategy is too common to have arisen later. Along with this, there must have been a minimum set of membrane transport proteins that could move ions and substrates into the cell (and out of the cell) without leaking too many protons. This cell must have had the ability to make many coenzymes and other special molecules like heme, flavins and iron sulfur centers. The informational molecule was probably DNA, with genes being transcribed to RNA and then to protein on ribosomes. A DNA polymerase and an RNA polymerase were present. Oxygen was not available, so the electron acceptor at the end of the electron transport chain may have been a sulfur compound or Fe^{3+}. This cell probably had the protein export machinery needed to construct a cell wall outside the plasma membrane. The cell could make its own lipids, though it is not clear whether it made ether lipids or ester lipids. Certainly it had some important biochemical pathways for purines, pyrimidines and probably all 20 amino acids. This last point is supported by the fact that some enzymes that are clearly present in all three domains can be aligned with some certainty. The aligned positions of these enzymes include highly conserved positions that represent all 20 amino acids. Therefore, all 20 were apparently present in the last common ancestor. It is not clear if it was photosynthetic or not, but the similarities between ancient stromatolites and modern stromatolites, that have photosynthetic cyanobacteria, suggest photosynthesis was probably present in the 3.5 billion year old stromatolites. If the common ancestor was not photosynthetic, then the split between bacteria and archaea had to predate 3.5 billion years ago. This is the idea favoured by Christian de Duve in his book *Vital Dust*. He argues that no archaebacteria have chlorophyll, and relatively few bacteria are photosynthetic. Therefore, chlorophyll and photosynthesis must have developed after archaea and bacteria split, but before 3.5 billion years ago, when cyanobacteria were probably photosynthetic.

Tree of life

The root of a phylogenetic tree is the location of the last common ancestor shared by all members of the tree. The root of a tree cannot be determined without an outgroup, a sequence that is related to the sequences of interest, but not a direct member of that group. For example, if you wanted to root the tree of mammalian ADP/ATP carriers, you could include a fungal ADP/ATP carrier as an outgroup. The point on the tree where the outgroup joins the other sequences is the root. When there are only three groups being considered, like archaea, bacteria and eukarya, you cannot root the tree, because there is no outgroup. Clever molecular evolutionists figured out a way to root this tree by making a tree with duplicated genes that are very old and are found in all the domains of life. This means that the duplication preceded the divergence of these three domains.

Two duplicated proteins that were used for this are the alpha and beta subunits of the vacuolar V-type and F1FO ATPases. This did not work too well, because there are many different V-type and F1FO ATPases, and there is some evidence that there might have been lateral gene transfer. Another protein set that does not have this criticism are the elongation factors EF-Tu and EF-G. The EF-G branch on the tree serves to root the Ef-Tu branch and vice versa. This was reported in the July 96 PNAS vol 93, 7749-7754. A similar result was found with isoleucyl tRNA synthetase sequences). The analysis of the EF-Tu and EF-G sequences strongly supports the bacteria/archaea split at the root of the tree. However, the frequently drawn branching of eukaryotes from archaea does not occur. Instead, eukaryotes branch from within the archaea. They appear to cluster with the crenarchaeota. The authors caution that more proteins need to be analyzed in this way and the data is not absolutely convincing. They also point out that other results have been seen with glutamine synthase, glutamate dehydrogenase and Hsp70 sequences.

Eukaryotes

First, eukaryotes have a nucleus. Neither archaea or bacteria have a nucleus, and it is very hard to imagine how such a structure would evolve. Fusion of two cells would offer an explanation of how a nucleus could be formed. If a gram negative bacterial cell engulfed an archael cell, it could wrap it in its outer membrane to form a structure that would be like a nucleus. If the archael cell lost its membrane and used the host membrane instead, this would form the nuclear envelope and the endoplasmic reticulum.

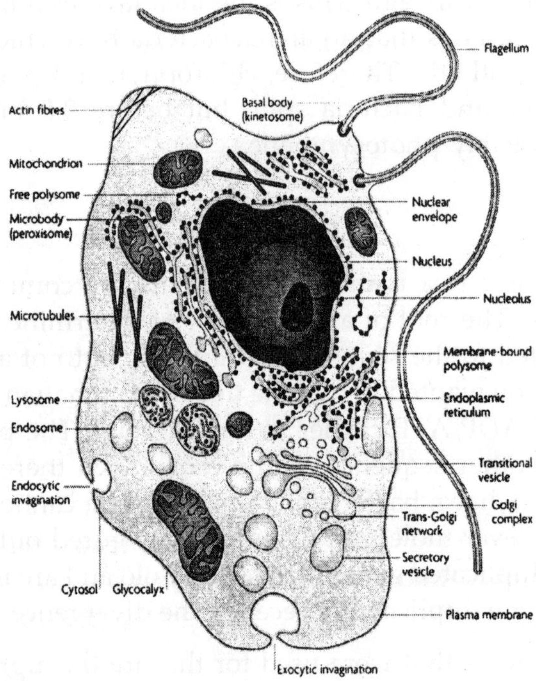

Figure 1. Eukaryotes

The hydrogen hypothesis is the newest alternative theory. This hypothesis claims that there was no benefit to the host from aerobic respiration brought in by the mitochondrial ancestor. Instead, the whole process was anaerobic between a host that was a methanogen that converted CO_2 and H_2 into methane like modern day archaeal methanogens. The symbiont would be a bacterium that could respire in aerobic conditions or ferment in anaerobic conditions producing CO_2 and H_2 as waste products. Since the waste products are the starting compounds for the methanogen, there would be an strong selection to associate. The theory then proposes movement of genes from the symbiont to the host for glycolysis enzymes and sugar transporters for fuel molecules for the symbiont, so the host can support the symbiont and completely enclose it from the environment. In an aerobic environment, the symbiont can respire, but it is not a benefit to the host until the ADP/ATP carrier can evolve to export the ATP to the cytosol. Loss of respiratory enzymes results in formation of a hydrogenosome organelle, a degenerate mitochondrion.

RNA

Several years ago Tom Cech discovered that RNA could splice introns out of itself (type I and type II introns are self splicing) without any protein being present. This meant that RNA had enzymatic activity. This changed our concept of the origin of life, by saying that proteins were not needed early in the evolution of life. A complete DNA to RNA to protein apparatus was not needed. DNA was not even needed. Everything could be done by RNA. The concept of life without DNA or protein is the RNA world. The RNA world is presented at a page by Leslie Orgel, an origin of life researcher. There are several key biochemical processes that depend on RNA components. These are listed below.

1) Ribosomes have RNA in their structure that is essential for function. In fact, it is beginning to look like the formation of peptide bonds is catalyzed by the RNA part of the ribosome and not the protein part.

2) Messenger RNA is a key ingredient in information transfer. Many viruses are RNA viruses that use RNA as the main informational macromolecule. RNA seems to have preceded DNA as the informational macromolecule for directing protein synthesis.

3) Protein export machinery of the endoplasmic reticulum binds a protein RNA complex called the signal recognition particle SRP. SRP contains an essential RNA called 7SL RNA.

4) Telomerase, required to maintain the ends of linear chromosomes has an RNA template that is needed to restore chromosome ends after replication.

5) RNAase P is a protein RNA complex needed for maturation of the 5' end of tRNAs It has an RNA subunit and a protein subunit. In bacteria, it is clear that the RNA subunit is responsible for enzymatic activity. The protein enhances activity by 20 fold.

6) snRNAs (small nucleolar RNAs) are needed to process mRNA for export to the cytosol, by removing introns. These are found in the spliceosome.

7) tRNAs appear to predate protein synthesis. The anticodon loop seems to be a later evolutionary development. Some tRNAs have specific catalytic roles, such as the formation of delta aminolevulinate in heme biosynthesis in plants.

8) Many coenzymes contain ribose, and not deoxyribose sugars. Ribozymes had so few useful structures to catalyze reactions that they needed many coenzymes to add the necessary functionality.

The nucleotide part of these coenzymes does not participate in the reactions. Instead, it helps the protein recognize and bind the coenzyme. Ribozymes also could bind these coenzymes more readily if they contained nucleotides that could hydrogen bond with the RNA nucleotides. This may be a reason why so many coenzymes have nucleotide parts: NAD, FAD, CoA, S-adenosylmethionine, cobalamin, the major energy currency of the cell, is a ribose containing RNA nucleotide. 10) Group I and group II introns can splice themselves. The RNA cuts and religates without any help from proteins in vitro. In vivo, proteins aid the reaction. 11) stRNA (small temporal RNAs) also called miRNAs for micro RNAs are only 21-24 nucleotides long, but they are conserved between human and worm and have a role in timing development. It is not known how far back in time they will be found. One feature that is not well conserved in all three domains is ribonucleotide reductase. Each domain of life has a very different enzyme that uses different cofactors to catalyze this important reaction. The reaction is the removal of the 2' OH group from ribonucleotides. The absence of a conserved protein sequence for this function suggested that this reaction was being catalyzed by an RNA ribozyme in the last common ancestor.

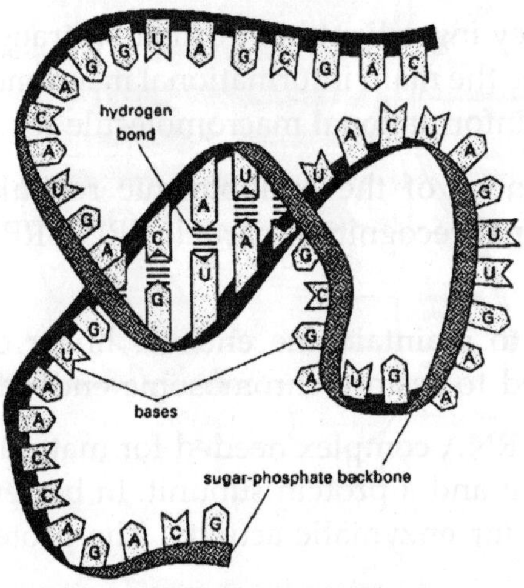

Figure 2. RNA structure

The argument follows that this ribozyme has since been replaced, but independently in all three lineages. This argument is now known to be untrue. Sequence similarity has been shown between the class I and class II enzymes (recently sequenced in a genome project from an Archaea). The crystal structure for a class III enzyme has been solved and the fold is a 10 stranded beta/alpha barrel as in the class I enzyme, even though the sequences are very different.

The class III enzyme has a glycyl radical instead of a tyrosine radical and it is poisoned by oxygen (it actually cuts off its own tail). The thought now is that the class I and class II forms evolved after oxygen arose in the atmosphere to make an oxygen stable radical that would not self destruct. The fact that there are de novo pathways for ribonucleotides but not for deoxyribonucleotides (they are only made from ribonucleotides) suggests that DNA came after RNA in evolution.

A related issue is the question of which came first DNA or protein. The mechanism is a rare and energetically costly one and it looks like it may be the only possible way to do this reaction, since no other enzyme type has been found to do it. The possibility that RNA could catalyze a free radical reaction seems unlikely, therefore protein was needed before ribonucleotide reductase could exist and therefore before DNA could be made.The lipids of modern archaea are ether linked lipids, while bacteria and eukarya have ester linked lipids. The archaea with the highest heat tolerance have ether linked lipids that are covalently coupled across the bilayer. This adds to their membrane stability at high temperatures. Christian de Duve suggests that the common ancestor was an ether linked species that had to switch to ester linkages to survive a drop in temperature.

The present day archaea have some members that can grow at 110 degrees, but the most thermophilic bacteria can only stand about 80 degrees. He suggests this is related to lipid composition. If the common model of branching is correct, then the transition to ester linked lipids had to occur twice, once in the bacteria and again in the eukarya. If the chimeric model of eukarya is correct, the gram negative bacteria that fused with archaea could have brought the ester linked lipid biosynthetic genes along. Murein is only found in bacteria not archaea. It is a cell wall component made of sugars and D and L amino acids. The presence of both D and L amino acids suggests an ancient origin for this material. Archaea may have lost the ability to make murein, however, it is possible that the synthesis of murein evolved after bacteria and archaea split.

Cholesterol is a eukaryotic lipid. Not all eukaryotes make it. Yeast make ergosterol instead and plants make cycloartenol derivatives, though palms do make cholesterol. There are reports that three bacteria can make partially demethylated lanosterol products, on the way to cholesterol. But, for the most part cholesterol biosynthesis seems to be a eukaryotic phenomenon.

Photosynthesis

Life in the early earth was anaerobic. In the history of photosynthesis, photosystem I must have developed first. PSI cannot split water to make O_2, so the first source of electrons for this pathway had to be mineral compounds. Eventually, PSII evolved from PSI and the ability to split water and form O_2 changed the planet. Here is a quote "The creation of a photosynthetic apparatus capable of splitting water into oxygen, protons and electrons was the pivotal innovation in the evolution of life on Earth. This event literally changed the face of the Earth." O_2 is the source of many toxic oxygen compounds such as superoxide anion, hydrogen peroxide and hydroxyl radical. These are known as reactive oxygen species (ROS). Cells had to protect themselves from these oxygen byproducts of photosynthesis. Of course, the cells that produced these compounds had to protect themselves first. This required the evolution of free radical scavengers like vitamins C and E, and enzymes like superoxide dismutase and catalase.

Oxygen released to the environment by photosynthesizing cyanobacteria would react with any oxidizable substance. The early oceans were believed to contain a lot of Fe^{2+} that could react with oxygen to form rust. There is evidence from 3.75 billion year old sedimentary deposits called banded-iron formations, that Fe^{2+} served as an oxygen sink. These deposits formed from 3.75 billion years to about 1.7 billion years. As these deposits began to lessen, oxygen began to appear in the atmosphere, starting about 2 billion years ago. It has remained pretty constant since 1.5 billion years ago. The inference is that iron in the oceans served as an oxygen trap for almost 2 billion years until it was all oxidized. As the iron became depleted, oxygen concentrations in the atmosphere could rise.

Organisms that were not O_2 producers either had to hide from O_2 or adapt. There are still anaerobes that hid from the toxic effects of O_2. The other organisms adapted and developed protections from oxygen. These organisms eventually learned how to exploit O_2 as the ultimate electron acceptor of their electron transport chains and gained a great advantage in the amount of energy that could be recovered in the form of ATP. The presence of oxygen also opened up the possibilities of oxygen chemistry that were not available before. Note that none of the amino acid biosynthesis pathways requires molecular oxygen, but many of the breakdown pathways do. You should also be aware that cholesterol biosynthesis requires oxygen to remove a 14 alpha methyl group from the cholesterol ring structure. This is needed to make cholesterol planar, which is an important feature of this critical eukaryotic lipid. Cholesterol cannot be made without oxygen and the cytochrome P450 CYP51.

Mitochondria and Chloroplasts

It is thought that the earliest eukaryotes had no mitochondria or chloroplasts, but they acquired these organelles by endosymbiosis. Today the most ancient branches on the eukaryotic tree all represent groups without mitochondria. The earliest eukaryotes seem to be the Archaezoans diplomonads (includes Giardia), and the trichomonads. The evidence is very strong that mitochondria were taken in as endosymbionts of the alpha proteobacteria. Phylogenetic trees of the rRNAs from bacteria and mitochondria show this plainly. Hydrogenosomes and peroxisomes may also be endosymbionts, but they do not contain a genome of their own today.

Sequences of the largest subunit of RNA polymerase II show that microsporidians are really related to fungi. Their deep branching on rRNA trees is artifactual. PNAS 96, 580-585, 1999 Microsporidia are related to Fungi:Evidence from the largest subunit of RNA polymerase II and other proteins. Microsporidians also contain a characteristic insertion in their EF1 alpha protein that is only seen in the fungal and animal clade (Opisthokonta). This strongly suggests they must belong to this clade.

Mitochondrial-like heat shock protein genes are found in the genomes of these organisms. Because of this, the use of Giardia as a model of the primitive eukaryote before mitochondria were acquired is no longer viable. In fact, there is no living model for the earliest eukaryote that existed before mitochondria. Recent work has shown that Giardia are part of a very early branch on the eukaryotic tree All eukaryotic alanyl tRNA synthetases in eukaryotes have been replaced by a bacterial version probably from the mitochondrial endosymbionts genome. This did not happen in Giardia where the synthetase is like an archaeal gene.

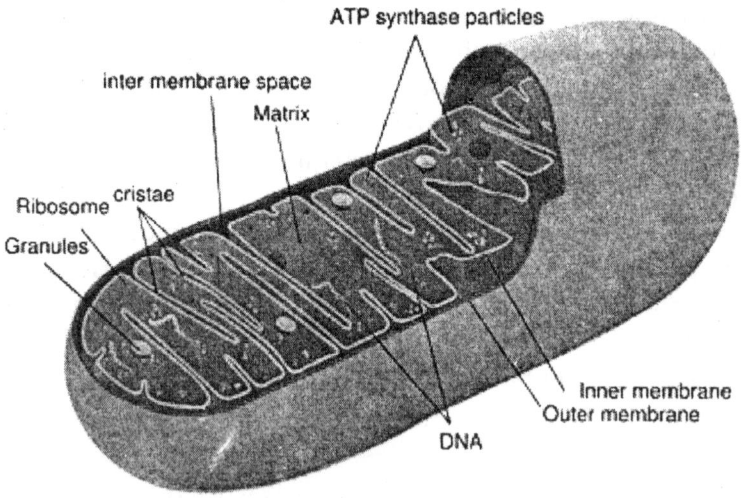

Figure 3. Mitochondria inner structure

Hydrogenosomes are organelles found only in eukaryotes that do not have mitochondria. These organelles are surrounded by a double membrane like mitochondria. They are the site of pyruvate fermentation to produce acetate, CO_2 and H_2. ATP is formed by substrate level phosphorylation, so these organelles resemble mitochondria in that they produce ATP. They do not have pyruvate dehydrogenase. Instead they use an enzyme called pyruvate ferredoxin oxidoreductase, and they have hydrogenase. These enzymes are not found in mitochondria, but they are seen in anaerobic bacteria. Hydrogenosomes do not have an electron transport chain and F1F0 ATPase. They do not perform oxidative phosphorylation. But they do have an ADP/ATP carrier to exchange ATP and ADP with the cytosol. Recently a ciliate (Nyctotherus ovalis) was found with a hydrogenosome that still has a genome of its own. The small subunit rRNA gene was PCRed out and found to match with other mitochondrial versions of this gene.

One set of proteins found in hydrogenosomes are the heat shock proteins Hsp70, Hsp60 and Hsp10. These are ideal sequences to use for phylogenetic analysis. As we saw in the first evolution lecture, Gupta And Golding used them to argue for a hybrid eukaryote genome. A sequence analysis of the hydrogenosome Hsp70 and Hsp60 proteins showed a signature sequence characteristic of these proteins in mitochondria and gram negative purple bacteria. Trees made with other available sequences of all three Hsp proteins placed each of them in the mitochondrial Hsp group. Since this happened in all three cases, the hydrogenosome Hsps and the mitochondrial Hsps appear to have a common bacterial ancestor. In anaerobic environments where these organisms live, it was not useful to retain the OXPHOS genes encoded in the mitochondrial genome, which may explain why most hydrogenosomes have no genome of their own. These organelles are apparently a degenerate version of the mitochondrial ancestor.Before the eukaryotic ancestor could engulf these bacteria, it would have to evolve into a phagocytic cell, with the ability to take a whole living bacterium inside itself. This is not possible with a rigid cell wall. Therefore, the precursor of the eukaryote host had to lose the ancestral cell wall. An example of symbiosis between amebas and a bacterium has been documented in the lab. Kwang Jeon of our own UT (at Knoxville) was studying amebas when he noticed a bacterial infection had killed most of the cells in one of his cultures. He isolated the bacterium and tested it on another group of amebas. Again it killed most of them, but some survived and had 40,000 bacteria living inside the ameba cells. He cultured the amebas for many generations and then tried an experiment where he removed the nucleus from these infected amebas and transplanted it to the original uninfected strain. The amebas died unless he "infected" them with the bacteria. In other words, the host had become dependent on the bacteria and needed them to survive.

Introns

Introns are found in eukaryotic genes. There are two opposing views on what the origin of introns might be. In one view, introns are ancient and were present in all genes. The amino acid coding regions of genes were made of small pieces of 15-20 codons each and these were all spliced together by removal of the introns. In this model, genetic diversity was assured by exon shuffling, that combined different exons to make a very large collection of proteins. This is the exon theory of genes. Since these types of introns are not in bacteria, the theory goes that bacteria are streamlined and have lost all their introns. Furthermore, the exons in this theory are supposed to code for units of protein structure, such as helices and beta strands.

The "introns late" theory says that the original common ancestor did not have introns and they only evolved in the eukaryote branch of the tree of life. This theory suggests that introns are placed pretty much at random into genes, and exons do not necessarily correspond to protein structural elements like helices and strands. Triose phosphate isomerase is an enzyme that was used to support the introns early hypothesis, because the chicken TPI gene has six introns and they all occur between structural elements in the protein. When additional sequences from a plant and a fungus were determined, additional introns were found. Five were in the same place in plants and animals suggesting they existed in the common ancestor to plants and animals. The total number of presumed introns was now 11, with different lineages losing different introns over time. One of the new exons was big and did not fit well with the theory that compact modules of protein were encoded in the exons.

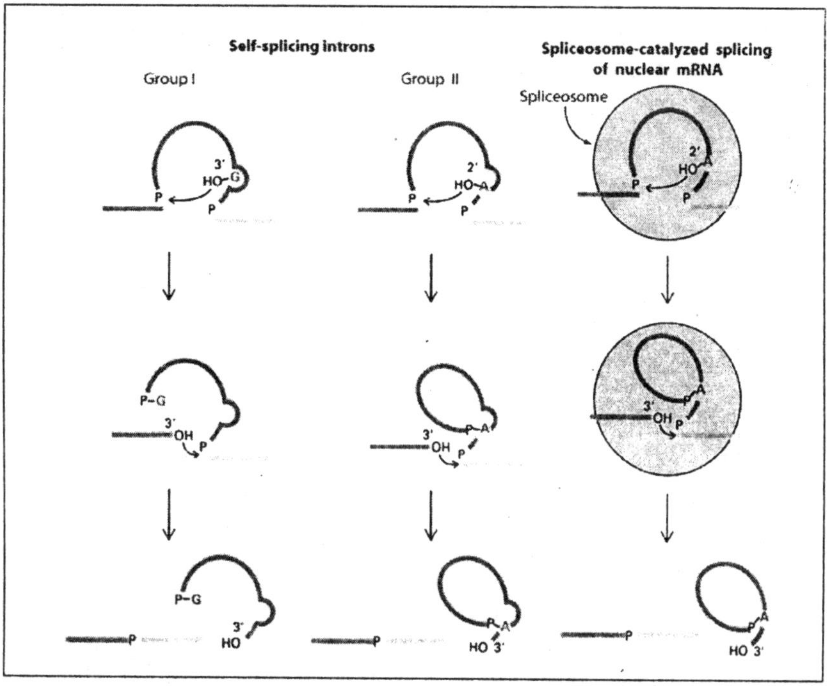

Figure 4. Introns

The issue was not solved yet, because more sequences were done from another insect, another fungi and C. elegans (nematode). These identified seven new intron positions, for a total of 21 introns and an average exon size of 11.2 codons. 12 introns only occur in one sequence, suggesting that all other lineages lost that intron. The exon theory then is getting to be very cumbersome. Additional sequences from more insects showed that a close relative of the Culex mosquito (Aedes mosquitoes) had the intron, but more distant relatives (Anopheles mosquitoes, flies and moths) did not. This is consistent with late insertion of the intron in an ancestor of the Culex and Aedes mosquitoes. Since 19 species are missing this intron, at least 10 independent losses of this intron would be required to fit the introns early model. Therefore, the introns early model seems to be wrong.

Sidney Brenner who is responsible for C. elegans as a model organism has used introns as a tool to follow evolution. If intron insertion or loss is a rare event, then the presence or absence of introns in a lineage should correlate with a single historic event when the intron was inserted (or lost). He has looked at seven introns from genes in the fugu fish but without the intron in man as a test case. By sampling many different species from all the major groups of fishes, he could see when the introns were inserted or lost. This is much better than looking at average numbers like percent sequence identity between genes, since those numbers reflect many individual mutations, that can occur at the same site more than once and can back mutate. There is probably no noise to the signal from intron insertion or loss.

Archaea and Eukarya

Archaea and eukarya share some distinctive features, including N-linked glycoproteins, absence of N-formylmethionine and introns in their tRNAs. N-linked glycoproteins are made in eukaryotes in the endoplasmic reticulum and the golgi. They are initially formed using a dolichol carrier lipid intermediate. Their presence in archaea suggests that dolichol is there and some membrane bound system similar to the N-linked biosynthetic machinery of the ER is also present. Bacteria use N- formylmethionine at the beginning of all their proteins. There is a special tRNA for this amino acid. Eukarya and archaea do not have N-formylmethionine.

Some eukaryotic tRNA genes have introns. In yeast, there are 262 tRNA genes, and about one third contain introns. Archaea also have introns in some of their tRNAs. Methanococcus jannaschii has 37 tRNA genes, with introns in a met and trp tRNA. The transcriptional apparatus of archaea is much more eukaryote-like than bacteria-like. This may reflect similarities in how DNA is packaged in the two domains. Methanococcus jannaschii has five histone genes, so the DNA may be packaged in nucleosomes as in eukaryotes. This may require a more complex transcription machinery to get at the DNA. Bacteria don't have histones and they have a much simpler transcriptional apparatus.Archaea and bacteria both have polycistronic operons, and some of these have the genes arranged in a similar order in both lineages. This implies that the operons existed in the common ancestor. Recently, there has been some evidence that C. elegans, a model organism for genome sequencing also has operons. This was unheard of in animals before these C. elegans operons were described.

Homeobox Genes

Humans, worms and flies don't look very similar and they do not go through the same developmental stages. Yet the genes that control their body shape and organization are related in sequence. These genes all share a common sequence called the homeobox. This 180 nucleotide sequence codes for 60 amino acids found in these proteins. The rest of the proteins may be very different, but this 60 amino acid piece is crucial for their function. The homeodomain is a helix turn helix DNA binding domain that recognizes a specific DNA sequence. The homeodomain targets the remainder of the protein to regulate the gene expression of any genes with the appropriate recognition sequence in their control regions. There are at least 50 homeobox genes in Drosophila. They fall into two main divisions, the complex superclass and the dispersed superclass. Those in the complex group are found in clusters, the dispersed group are solo genes. One subset of these genes are called homeotic selector genes. In Drosophila, there are 8 genes arranged in a series along 650,000 base pairs of DNA. This whole region is called the HOM complex. There are two smaller subsets of these genes in the HOM complex, the antennapedia complex (5 genes) and the bithorax complex (3 genes). Other insects have these genes all in one complex, so it looks as though the HOM complex became split in Drosophila.

Mutations in the 8 genes of the HOM complex cause large scale mutations in flies. A mutation in bithorax causes a fly to have an extra set of wings. Mutation in antennapedia causes a leg to grow where an antenna should be. These genes are not master switches for making wings or legs, but they specify position in the fly's body. The order of the genes on the chromosome is the same as the order of segments in the fly's body where they are expressed. The left most gene is expressed in the head, the right most gene is expressed in the abdomen. When a gene is deleted or mutated, the segment where it is normally expressed cannot tell where it is because its position clue is gone, so it behaves like the closest segment to it. That is why a bithorax mutation causes an extra set of wings. The segments adjacent to the bithorax segment dictated what should be made.

An amazing fact is that these HOM genes have clear homologs in vertebrates. These are called hox gene clusters. Mice have four hox gene clusters on four different chromosomes. These are called HoxA ,B, C and D. HoxB has all the same genes as HOM plus one more. They are in exactly the same order. The other three segments are missing some of the HOM genes, but they have some extra homeobox genes not in the HOM cluster. The HOM cluster seems to have arisen by gene duplication of a single homeobox gene long ago. This cluster then was duplicated in total four times in the lineage of vertebrates . Some additional gene duplication and deletion resulted in the present day set of Hox genes in mammals. These genes specify position in mouse embryos, just like they did in flies. They seem to have a similar function to the HOM cluster in flies, except it is more complicated in mammals because there are four clusters. Two sets of hox gene clusters are expressed in limb buds in perpendicular directions.

The gene products from one cluster are expressed along a left to right axis in the limb bud(HoxD) and the other gene cluster is expressed top to bottom in the same bud(HoxA). This creates a checkerboard pattern that makes each position in the limb bud unique, like the elements of a mathematical array that are described by x, y and z coordinates. If a single gradient in the fly can specify the development of different symmetrical segments, like head, thorax and abdomen, then a dual gradient in the limb bud can specify the development of asymmetry in the limbs, things like the bones and muscles of the hand, the layout of nerves and blood vessels, what is to be skin and fingernails.The HoxA cluster in mouse has 11 genes, Drosophila has eight genes in the HOM cluster.

HoxA has added three extra genes. Probably, if one looks back at simpler organisms there will be some that have fewer homeotic genes in these clusters, or fewer clusters. The Annual Review of Biochemistry 1994 has an article on homeodomain proteins. There, evidence is cited for one hox cluster in acorn worms (a hemichordate), two hox clusters in amphioxus (a cephalochordate) and three (or 4) in lamprey (a primitive vertebrate).

Another homeobox gene in Drosophila is eyeless. This gene appears to be a master switch gene that turns on eye formation. If eyeless is expressed in tissues where it normally would not be active, whole functional eyes form. These may be on the end of antenna, on the wings or on the legs. This gene has homologs in mouse and man (aniridia) that also affect the formation of eyes. In fact, the mouse gene can substitute in Drosophila for the eyeless gene. This means that eye formation is controlled by a gene that evolved before eyes evolved. The common ancestor apparently had a light sensitive tissue that later evolved into different types of eyes in insects and vertebrates. Eyeless controlled the development of that light sensitive tissue and it has continued in that role for at least 600 million years.

A recently discovered gene called Manx is not a homeobox gene, but it is a zinc-finger transcription factor and it is another candidate for a master switch gene.The Manx gene is found in tunicates, a type of primitive chordate. These organisms start life as a tadpole like creature with a tail and notochord. During maturation, they lose their tail and become sessile on the sea bottom. Manx is the gene that controls the tail formation. If it is mutated, the tail never forms. This was demonstrated by William Jeffery and Billie J. Swalla, who found two closely related tunicates, one with a tail and one without. They bred the two and the hybrid had a small tail, suggesting that a single gene was responsible and one functional copy could turn on the pathway.

The gene was identified and its expression was blocked by antisense RNA in the hybrid embryo. When this was done, the tail did not form. They are now looking for homologs in more complex vertebrates. Manx is similar to eyeless, in that it turns on a whole developmental program to form a tail. Of course the hunt is on to see what gene might control Manx and which genes lie downstream of Manx to effect tail development. A second way to bring about macroevolution is polyploidy. Xenopus laevis has twice the DNA of Xenopus tropicalis, by a genome duplication. This gives Xenopus laevis a lot of DNA to experiment with and try out new functions for old duplicated genes. One consequence of doubling the number of genes is an increase in size. Xenopus laevis is much larger than Xenopus tropicalis, perhaps due to a gene dosage effect. As we mentioned above, the fish seem to have undergone a genome duplication also. Plants are often tetraploid or hexaploid, again giving evolution a lot of material to work with.

Polymorphisms

Polymorphisms are fixed sequence differences in a population that make up more than

1% of the population. The HLA locus in humans corresponds to the MHC locus in mice and other vertebrates. This is a highly polymorphic region with about 100 genes. One of these genes is the DRB1 gene. 59 alleles exist in humans and 60 non-human primate sequences have been determined. A tree of these sequences gives an estimate of the time for a last common ancestor of about 60 million years. It is important to point out here, that these 59 alleles are different sequence variants of the same gene in humans. These do not represent different genes.

To carry that many polymorphisms in a population, the absolute minimum number of individuals would be 30, one diploid person for every two alleles, and they would all have to be heterozygotes, each with a different allele. This situation is very unlikely. Six million years ago, there were 32 lineages of the DRB1 gene, with a minimum population to carry this number of alleles being 16. Again the actual population would have to be much greater than that. There is a theory dealing with polymorphisms and population size. This is called coalescence theory. If the time of coalescence is known, and the number of genes is known, then this theory will predict what the population size must be for this to happen. The results of simulations show that for 60 alleles to persist for 1.7 million years (time of humans as Homo sapiens) the population size would have to be about 100,000. If it was less, many of the alleles would become lost over that many generations. The numbers are not incompatible with a mitochondrial Eve hypothesis, because mitochondrial Eve is only considering the inheritance of one small piece of DNA, equivalent to a single gene. It must be true that a single woman of about 200,000 years ago is the mother of all of our mitochondrial DNA, but it is not true that she is the mother of all our other genes.

Genetics of Evolution

The central argument of Darwin's theory of evolution starts from the existence of hereditary variation. Experience with animal and plant breeding demonstrates that variations can be developed that are useful to humans. So, reasoned Darwin, variations must occur in nature that are favourable or useful in some way to the organism itself in the struggle for existence. Favourable variations are ones that increase chances for survival and procreation. These advantageous variations are preserved and multiplied from generation to generation at the expense of less advantageous ones in the process known as natural selection. The outcome of the process is an organism that is well adapted to its environment, and evolution often occurs as a consequence.

Biological evolution is the process of change and diversification of living things over time, and it affects all aspects of their lives—morphology, physiology, behaviour, and ecology. Underlying these changes are changes in the hereditary materials. Hence, in genetic terms, evolution consists of changes in the organism's hereditary makeup. Natural selection, then, can be defined as the differential reproduction of alternative hereditary variants, determined by the fact that some variants increase the likelihood that the organisms having them will survive and reproduce more successfully than will organisms carrying alternative variants. Selection may be due to differences in survival, in fertility, in rate of development, in mating success, or in any other aspect of the life cycle. All of these differences can be incorporated under the term 'differential reproduction', because all result in natural selection to the extent that they affect the number of progeny an organism leaves.

Darwin explained that competition for limited resources results in the survival of the most effective competitors. But natural selection may occur not only as a result of competition but also as an effect of some aspect of the physical environment, such as inclement weather. Moreover, natural selection would occur even if all the members of a population died at the same age, if some of them produced more offspring than others. Natural selection is quantified by a measure called Darwinian fitness, or relative fitness. Fitness in this sense is the relative probability that a hereditary characteristic will be reproduced; that is, the degree of fitness is a measure of the reproductive efficiency of the characteristic.

Evolution can be seen as a two-step process. First, hereditary variation takes place; second, selection occurs of those genetic variants that will be passed on most effectively to the following generations. Hereditary variation also entails two mechanisms: the spontaneous mutation of one variant to another, and the sexual process that recombines those variants to form a multitude of variations.

The gene pool of a species is the sum total of all of the genes and combinations of genes that occur in all organisms of the same species. The necessity of hereditary variation for evolutionary change to occur can be understood in terms of the gene pool. Assume, for instance, that at the gene locus that codes for the human MN blood groups, there is no variation; only the *M* form exists in all individuals. Evolution of the MN blood groups cannot take place in such a population, since the allelic frequencies have no opportunity to change from generation to generation. On the other hand, in populations in which both forms *M* and *N* are present, evolutionary change is possible.

The more genetic variation that exists in a population, the greater the opportunity for evolution to occur. As the number of genes that are variable increases and as the number of forms of each gene becomes greater, the likelihood that some forms will change in frequency at the expense of their alternates grows. The British geneticist R. A. Fisher mathematically demonstrated a direct correlation between the amount of genetic variation in a population and the rate of evolutionary change by natural selection. This demonstration is embodied in his fundamental theorem of natural selection: "The rate of increase in fitness of any organism at any time is equal to its genetic variance in fitness at that time." This theorem has been confirmed experimentally. Because a population's potential for evolving is determined by its genetic variation, evolutionists are interested in discovering the extent of such variation in natural populations. Techniques for determining genetic variation have been used to investigate numerous species of plants and animals. Typically, insects and other invertebrates are more varied genetically than mammals and other vertebrates; and plants bred by outcrossing exhibit more variation than those bred by self-pollination. But the amount of genetic variation is in any case astounding. Every individual represents a unique genetic configuration that will never be repeated again. The enormous reservoir of genetic variation in natural populations provides virtually unlimited opportunities for evolutionary change in response to environmental constraints and the needs of the organisms, beyond the new variations that arise every generation by the process of mutation.

Variation and Mutation

There are more than two million known species, which are widely diverse in size, shape, and ways of life, as well as in the DNA sequences that contain their genetic information. What has produced the pervasive genetic variation within natural populations and the genetic differences among species? The information encoded in the nucleotide sequence of DNA is, as a rule, faithfully reproduced during replication, so that each replication results in two DNA molecules that are identical to each other and to the parent molecule. But, occasionally 'mistakes', or mutations, occur in the DNA molecule during replication, so that daughter cells differ from the parent cells in at least one of the letters in the DNA sequence. A mutation first appears on a single cell of an organism, but it is passed on to all cells descended from the first. Mutations can be classified into two categories: gene, or point, mutations, which affect only a few letters (nucleotides) within a gene; and chromosomal mutations, which either change the number of chromosomes or change the number or arrangement of genes on a chromosome.

Gene mutations can occur spontaneously; that is, without being intentionally caused by humans. They can also be artificially induced by ultraviolet light, X irradiation, and other high-frequency types of radiation, as well as by exposure to certain mutagenic chemicals, such as mustard gas. The consequences of gene mutations may range from negligible to lethal. Some have a small or undetectable effect on the organism's ability to survive and reproduce, because no essential biological functions are altered. But when the active site of an enzyme or some other essential function is affected, the impact may be severe. Newly arisen mutations are more likely to be harmful than beneficial to their carriers, because mutations are random events with respect to adaptation; that is, their occurrence is independent of any possible consequences. Harmful mutations are eliminated or kept in check by natural selection. Occasionally, however, a new mutation may increase the organism's adaptation. The probability of such an event's happening is greater when organisms colonise a new territory or when environmental changes confront a population with new challenges.

In these situations, the established adaptation of a population is less than optimal, and there is greater opportunity for new mutations to be better adaptive. This is so because the consequences of mutations depend on the environment. Increased melanin pigmentation may be advantageous to inhabitants of tropical Africa, where dark skin protects them from the sun's ultraviolet radiation; but it is not beneficial in Scandinavia, where the intensity of sunlight is low and light skin facilitates the synthesis of vitamin D. Mutation rates are low, but new mutants appear continuously in nature, because there are many individuals in every species and many genes in every individual. The process of mutation provides each generation with many new genetic variations. More important yet is the storage of variation, the results of past mutations, in each organism. Thus, it is not surprising to see that when new environmental challenges arise, species are able to adapt to them. More than 200 insect and rodent species, for example, have developed resistance to the pesticide DDT in different parts of the world where spraying has been intense. Although the insects had never before encountered this synthetic compound, they adapted to it rapidly by means of mutations that allowed them to survive in its presence. Similarly, many species of moths and butterflies in industrialised regions have shown an increase in the frequency of individuals with dark wings in response to environmental pollution, an adaptation known as industrial melanism. The examples can be multiplied at will.

Genetic Equilibrium and Genetic Change

The genetic variation present in natural populations of organisms is sorted out in new ways in each generation by the process of sexual reproduction. But heredity by itself does not change gene frequencies. This principle is formally stated by the Hardy–Weinberg law, an algebraic equation that describes the genetic equilibrium in a population. In the simplest instance, there are two alleles, *A* and *a*, at a gene locus, so that three genotypes are possible, *AA*, *Aa*, and *aa*. If the frequencies of the two alleles in the population are p and q, respectively, the equilibrium frequencies of the three genotypes are given by $(p + q)^2 = p^2 + 2\,p\,q + q^2$ for *AA*, *Aa*, and *aa*, respectively. The genotype equilibrium frequencies for any number of alleles are derived in the same way.

The Hardy–Weinberg law assumes that gene frequencies remain constant from generation to generation—that there is no gene mutation or natural selection and that populations are very large. But these assumptions are not correct; indeed, if they were, evolution could not occur. Why, then, is the Hardy–Weinberg law significant if its assumptions do not hold true in nature? The answer is that the Hardy–Weinberg law plays in evolutionary studies a role similar to that of Newton's first law of motion in mechanics. Newton's first law says that a body not acted upon by a net external force remains at rest or maintains a constant velocity. In fact, there are always external forces acting upon physical objects (gravity, for example), but the first law provides the starting point for the application of other laws. Similarly, organisms are subject to mutation, selection, and other processes that change gene frequencies, and the effects of these processes are calculated by using the Hardy–Weinberg law as the starting point. There are four processes of gene frequency change: mutation, migration, drift, and natural selection. The allelic variations that make evolution possible are generated by the process of mutation; but new mutations change gene frequencies very slowly, since mutation rates are low. Moreover, gene mutations are reversible. Changes in gene frequencies due to mutation occur, therefore, very slowly. In any case, allelic frequencies usually are not in mutational equilibrium, because some alleles are favored over others by natural selection. The equilibrium frequencies are then decided by the interaction between mutation and selection, with selection usually having the greater consequence.

Migration, or gene flow, takes place when individuals migrate from one population to another and interbreed with its members. The genetic makeup of populations changes locally whenever different populations intermingle. In general, the greater the difference in allele frequencies between the resident and the migrant individuals and the larger the number of migrants, the greater effect the migrants have in changing the genetic constitution of the resident population.Moreover, gene frequencies can change from one generation to another by a process of pure chance known as genetic drift. This occurs because populations are finite in numbers, and thus the frequency of a gene may change in the following generation by accidents of sampling, just as it is possible to get more or fewer than 50 heads in 100 throws of a coin simply by chance. The magnitude of the gene frequency changes due to genetic drift is inversely related to the size of the population: the larger the number of reproducing individuals, the smaller the effects of genetic drift.

The effects of genetic drift in changing gene frequencies from one generation to the next are small in most natural populations, which generally consist of thousands of reproducing individuals. The effects over many generations are more important. Genetic drift can have important evolutionary consequences when a new population becomes established by only a few individuals, as in the colonisation of islands and lakes. This is one reason why species in neighboring islands, such as those in the Hawaiian archipelago, are often more heterogeneous than species in comparable continental areas adjacent to one another.

Natural selection

The phrase 'natural selection' was used by Darwin to refer to any reproductive bias favouring some hereditary variants over others. He proposed that natural selection promotes the adaptation of organisms to the environments in which they live because the organisms carrying such useful variants would leave more descendants than those lacking them. The modern concept of natural selection derives directly from Darwin's but is defined precisely in mathematical terms as a statistical bias favouring some genetic variants over their alternates (the measure to quantify natural selection is called fitness).

Hereditary variants, favourable or not to the organisms, arise by mutation. Unfavourable ones are eventually eliminated by natural selection; their carriers leave no descendants or leave fewer than those carrying alternative variants. Favourable mutations accumulate over the generations. The process continues indefinitely because the environments that organisms live in are forever changing. Environments change physically—in their climate, physical configuration, and so on—but also biologically, because the predators, parasites, and competitors with which an organism interacts are themselves evolving.

If mutation, migration, and drift were the only processes of evolutionary change, the organisation of living things would gradually disintegrate, because they are random processes with respect to adaptation. These three processes change gene frequencies without regard for the consequences that such changes may have on the ability of the organisms to survive and reproduce. The effects of such processes alone would be analogous to those of a mechanic who changed parts in an automobile engine at random, with no regard for the role of the parts in the engine. Natural selection keeps the disorganising effects of mutation and other processes in check, because it multiplies beneficial mutations and eliminates harmful ones. But natural selection accounts not only for the preservation and improvement of the organisation of living beings but also for their diversity. In different localities or in different circumstances, natural selection favours different traits, precisely those that make the organisms well adapted to their particular circumstances and ways of life.

The effects of natural selection can be studied by measuring the ensuing changes in gene frequencies; but they can also be explored by examining changes on the observable characteristics—or phenotypes—of individuals in a population. Distribution scales of phenotypic traits such as height, weight, number of progeny, or longevity typically show greater numbers of individuals with intermediate values and fewer and fewer toward the extremes (the so-called normal distribution). When individuals with intermediate phenotypes are favored and extreme phenotypes are selected against, the selection is said to be stabilising. The range and distribution of phenotypes then remains approximately the same from one generation to another. Stabilising selection is very common. The individuals that survive and reproduce more successfully are those that have intermediate phenotypic values. Mortality among newborn infants, for example, is highest when they are either very small or very large; infants of intermediate size have a greater chance of surviving.

But the distribution of phenotypes in a population sometimes changes systematically in a particular direction. The physical and biological aspects of the environment are continuously changing, and over long periods of time the changes may be substantial. The climate and even the configuration of the land or water vary incessantly. Changes also take place in the biotic conditions; that is, in the other organisms present, whether predators, prey, parasites, or competitors. Genetic changes occur as a consequence, because the genotypic fitnesses may be shifted so that different sets of variants are favored. The opportunity for directional selection also arises when organisms colonise new environments where the conditions are different from those of their original habitat. The process of directional selection often takes place in spurts. The replacement of one genetic constitution for another changes the genotypic fitnesses of genes for other traits, which in turn stimulates additional changes, and so on in a cascade of consequences.

The nearly universal success of artificial selection and the rapid response of natural populations to new environmental challenges are evidence that existing variation provides the necessary materials for directional selection, as Darwin already explained. More generally, human actions have been an important stimulus to this type of selection. Mankind transforms the environments of many organisms, which rapidly respond to the new environmental challenges through directional selection. Well known instances are the many cases of insect resistance to pesticides, synthetic substances not present in the natural environment. Whenever a new insecticide is first applied to control a pest, the results are encouraging, because a small amount of the insecticide is sufficient to bring the pest organism under control. As time passes, however, the amount required to achieve a certain level of control must be increased again and again, until finally it becomes ineffective or economically impractical. This occurs because organisms become resistant to the pesticide through directional selection. Resistance of the housefly, *Musca domestica,* to DDT was first reported in 1947. Resistance to one or more pesticides has now been recorded in more than 100 species of insects.

Sustained directional selection leads to major changes in morphology and ways of life over geologic time. Evolutionary changes that persist in a more-or-less continuous fashion over long periods of time are known as evolutionary trends. Directional evolutionary changes increased the cranial capacity of the human lineage from *Australopithecus afarensis,* human ancestors who lived four million years ago, with a small brain weighing somewhat less than one pound, to *Homo sapiens,* modern humans with a brain three and a half times as large. The more-or-less gradual increase in size during the evolution of the horse family from 50 million years ago to modern times is another of the many well studied examples of directional selection.

Two or more divergent traits in an environment may be favored simultaneously, which is called diversifying selection. No natural environment is homogeneous; rather, the environment of any plant or animal population is a mosaic consisting of more-or-less dissimilar subenvironments. There is heterogeneity with respect to climate, food resources, and living space. Also, the heterogeneity may be temporal, with change occurring over time, as well as spatial, with dissimilarity found in different areas. Species cope with environmental heterogeneity in diverse ways. One strategy is the selection of a generalist genotype that is well adapted to all of the subenvironments encountered by the species. Another strategy is genetic polymorphism, the selection of a diversified gene pool that yields different genetic makeups, each adapted to a specific subenvironment.

Sexual Selection and Kin Selection

One important factor in reproduction is mutual attraction between the sexes. The males and females of many animal species are fairly similar in size and shape except for the sexual organs and secondary sexual characteristics such as the mammary glands of female mammals. There are, however, species in which the sexes exhibit striking dimorphism. Particularly in birds and mammals, the males are often larger and stronger, more brightly colored, or endowed with conspicuous adornments. But bright colours make animals more visible to predators; for example, the long plumage of peacocks and birds of paradise and the enormous antlers of aged male deer are cumbersome loads at best. Darwin knew that natural selection could not be expected to favour the evolution of disadvantageous traits, and he was able to offer a solution to this problem. He proposed that such traits arise by 'sexual selection', which "depends not on a struggle for existence in relation to other organic beings or to external conditions, but on a struggle between the individuals of one sex, generally the males, for the possession of the other sex." Thus, the colored plumage of the males in some bird species makes them more attractive to their females, which more than compensates for their increased visibility to potential predators.

The apparent altruistic behaviour of many animals is, like some manifestations of sexual selection, a trait that at first seems incompatible with the theory of natural selection. Altruism is a form of behaviour that benefits other individuals at the expense of the one that performs the action; the fitness of the altruist is diminished by its behaviour, whereas individuals that act selfishly benefit from it at no cost to themselves. Accordingly, it might be expected that natural selection would foster the development of selfish behaviour and eliminate altruism. The beneficiaries of altruistic behaviour are usually relatives. They all carry the same genes, including the genes that promote altruistic behaviour. Altruism may evolve by kin selection, which is simply a type of natural selection in which relatives are taken into consideration when evaluating an individual's fitness.

Kin selection is explained as follows. Natural selection favours genes that increase the reproductive success of their carriers, but it is not necessary that all individuals with a given genetic makeup have higher reproductive success. It suffices that carriers of the genotype reproduce more successfully on average than those possessing alternative genotypes. A parent shares half its genes with each progeny, so a gene that promotes parental altruism is favored by selection if the behaviour's cost to the parent is less than half its average benefits to the progeny. Such a gene will be more likely to increase in frequency through the generations than an alternative gene that does not promote parental care. The parent spends some energy caring for the progeny because it increases the reproductive success of the parent's genes. But kin selection extends beyond the relationship between parents and their offspring. It facilitates the development of altruistic behaviour when the energy invested, or the risk incurred, by an individual is compensated in excess by the benefits ensuing to relatives.In many species of primates (as well as in other animals), altruism also occurs among unrelated individuals when the behaviour is reciprocal and the altruist's costs are smaller than the benefits to the recipient. This reciprocal altruism is found, for example, in the mutual grooming of chimpanzees as they clean each other of lice and other pests. Another example appears in flocks of birds that post sentinels to warn of danger. One crow sitting in a tree watching for predators, while the rest of the flock forages, incurs a small loss by not feeding, but this is well compensated by the protection it receives when it itself forages and others of the flock stand guard.

Origin of Species

Darwin sought to explain the splendid diversity of the living world: thousands of organisms of the most diverse kinds, from lowly worms to spectacular birds of paradise, from yeasts and molds to oaks and orchids. His *Origin of Species* is a sustained argument showing that the diversity of organisms and their characteristics can be explained as the result of natural processes. As Darwin noted, different species may come about as the result of gradual adaptation to environments that are continuously changing in time and differ from place to place. Natural selection favours different characteristics in different situations.

In everyday experience we identify different kinds of organisms by their appearance. Everyone knows that people belong to the human species and are different from cats and dogs, which in turn are different from each other. There are differences among people, as well as among cats and dogs; but individuals of the same species are considerably more similar among themselves than they are to individuals of other species. But there is more to it than that: a bulldog, a terrier, and a golden retriever are very different in appearance, but they are all dogs because they can interbreed. People can also interbreed with one another, and so can cats, but people cannot interbreed with dogs or cats, nor these with each other. It is, then, clear that although species are usually identified by appearance, there is something basic, of great biological significance, behind similarity of appearance; namely, that individuals of a species are able to interbreed with one another but not with members of other species. This is expressed in the following definition: species are groups of interbreeding natural populations that are reproductively isolated from other such groups.

The ability to interbreed is of great evolutionary importance, because it determines that species are independent evolutionary units. Genetic changes originate in single individuals; they can spread by natural selection to all members of the species but not to individuals of other species. Thus, individuals of a species share a common gene pool that is not shared by individuals of other species, because they are reproductively isolated. Although the criterion for deciding whether individuals belong to the same species is clear, there may be ambiguity in practice for two reasons. One is lack of knowledge: it may not be known for certain whether individuals living in different sites below to the same species, because it is not known whether they can naturally interbreed. The other reason for ambiguity is rooted in the nature of evolution as a gradual process. Two geographically separate populations that at one time were members of the same species later may have diverged into two different species. Since the process is gradual, there is no particular point at which it is possible to say that the two populations have become two different species. It is an interesting curiosity that some antievolutionists have referred to the existence of species intermediates as evidence against evolution; quite the contrary, such intermediates are precisely expected.

A similar kind of ambiguity arises with respect to organisms living at different times. There is no way to test whether or not today's humans could interbreed with those who lived thousands of years ago. It seems reasonable that living people, or living cats, would be able to interbreed with people, or cats, exactly like those that lived a few generations earlier. But what about the ancestors removed by one thousand or one million generations? The ancestors of modern humans that lived one million years ago (about 50 thousand generations) are classified in the species *Homo erectus*, whereas present-day humans are classified in a different species, *Homo sapiens*, because those ancestors were quite different from us in appearance and thus it seems reasonable to conclude that interbreeding could not have occurred with modern-like humans. But there is no exact time at which *Homo erectus* became *Homo sapiens*. It would not be appropriate to classify remote human ancestors and modern humans in the same species just because the changes from one generation to the next are small. It is useful to distinguish between the two groups by means of different species names, just as it is useful to give different names to childhood, adolescence, and adulthood, even though there is no one moment at which an individual passes from one to the next. Biologists distinguish species in organisms that lived at different times by means of a commonsense rule: if two organisms differ from each other about as much as two living individuals belonging to two different species differ today, they are classified into separate species and given different names.Bacteria and blue-green algae do not reproduce sexually, but by fission. Organisms that lack sexual reproduction are classified into different species according to criteria such as external morphology, chemical and physiological properties, and genetic constitution.

Adaptive radiation

The geographic separation of populations derived from common ancestors may continue long enough that the populations become completely differentiated species before coming together again. As the separated populations continue evolving independently, morphological differences may arise. Examples of adaptive radiation are common in archipelagos far removed from the mainland. The Galapagos Islands are about 600 miles off the west coast of South America. When Darwin arrived there in 1835, he discovered many species not found anywhere else in the world –for example, 14 species of finch (known as Darwin's finches). These passerine birds have adapted to a diversity of habitats and diets, some feeding mostly on plants, others exclusively on insects. The various shapes of their bills are clearly adapted to probing, grasping, biting, or crushing—the diverse ways in which these different Galapagos species obtain their food. The explanation for such diversity (which is not found in finches from the continental mainland) is that the ancestor of Galapagos finches arrived in the islands before other kinds of birds and encountered an abundance of unoccupied ecological opportunities. The finches underwent adaptive radiation, evolving a variety of species with ways of life capable of exploiting niches that in continental faunas are exploited by different kinds of birds.

Striking examples of adaptive radiation occur in the Hawaiian Islands. The archipelago consists of several volcanic islands, ranging from less than one million to more than ten million years in age, far away from any continent or other large islands. An astounding number of plant and animal species of certain kinds exist in the islands while many other kinds are lacking. Among the species that have evolved in the islands, there are about two dozen (about one-third of them now extinct) of honeycreepers, birds of the family Drepanididae, all derived from a single immigrant form. In fact, all but one of Hawaii's 71 native bird species are endemic; that is, they have evolved there and are found nowhere else. More than 90% of the native species of the hundreds of flowering plants, land mollusks, and insects in Hawaii are also endemic, as are two-thirds of the 168 species of ferns. About one-fourth of the world's total number of known species of *Drosophila* flies (more than 500) are native Hawaiian species. The species of *Drosophila* in Hawaii have diverged by adaptive radiation from one or a few colonisers, which encountered an assortment of ecological opportunities that in other lands are occupied by different groups of flies or insects.

Rapid or quantum speciation

Examples of rapid speciation are also known in many organisms. Instances of rapid speciation are sometimes called quantum or saltational speciation. An important form of quantum speciation occurs as a result of polyploidy, which is the multiplication of entire sets of chromosomes. This can happen in a single generation, for example, if meiosis fails so that an individual's gametes have two sets of chromosomes, rather than only one. If a male and a female gamete, each with two sets of chromosomes, combine, the resulting individual will have four sets, rather than two sets, of chromosomes. A typical (diploid) organism carries in the nucleus of each cell two sets of chromosomes, one inherited from each parent; a polyploid organism has several sets of chromosomes. Many cultivated plants are polyploid: bananas have three sets of chromosomes, potatoes have four, bread wheat has six, some strawberries have eight. All major groups of plants have natural polyploid species, but they are most common among flowering plants (angiosperms), of which about 47% are polyploids.In animals, polyploidy is relatively rare because it disrupts the balance between chromosomes involved in the determination of sex. But polyploid species are found in hermaphroditic animals (individuals having both male and female organs), which include snails and earthworms, as well as in forms with parthenogenetic females (which produce viable progeny without fertilisation), such as some beetles, sow bugs, goldfish, and salamanders.

References

Bendall, D. S., *Evolution From Molecules to Man*, Cambridge, Cambridge University Press, 1983.

Crow, James F. & Motoo Kimura, *Introduction to Population Genetics Theory*, Edina, Minn.: Burgess Publishing Company, 1970.

Cowen, Richard, *History of Life*, Boston: Blackwell Scientific, 2004.

Eldredge, Niles, *Macroevolutionary Dynamics*, New York: McGraw-Hill, 1989.

Fitch, W. M., "Molecular Evolutionary Clocks, *in* Ayala, F. J., ed., *Molecular Evolution: Sunderland*, Mass., Sinauer, 1976.

Futuyma, Douglas J., *Evolutionary Biology*, Sunderland, Mass.: Sinauer Associates, 1997.

Graur, Dan & Wen-Hsiung Li, *Fundamentals of Molecular Evolution*, Sunderland, Mass.: Sinauer Associates, 2000.

Gillespie, John H., *The Causes of Molecular Evolution*, New York: Oxford Univ. Press, 1997.

Glover, D. M., and Hames, B. D., *Genes and Embryos*, New York, Oxford University Press, 1989.

Golding, Brian, ed., *Non-Neutral Evolution*, Boston: Chapman and Hall, 1994.

Hartl, Daniel L. Andrew G. Clark, *Principles of Population Genetics*, Sunderland, Mass.: Sinauer Associates, 1997.

Kimura, Motoo, *The Neutral Theory of Molecular Evolution*, Cambridge, U.K.: Cambridge Univ. Press, 1983.

Lewontin, Richard C., *The Genetic Basis of Evolutionary Change*, New York: Columbia Univ. Press, 1974.

Provine, William B., *The Origins of Theoretical Population Genetics*, Chicago: Univ. of Chicago Press, 2001.

6

Molecular Biology of Enzymes

Enzymes are biomolecules that catalyze chemical reactions. Nearly all known enzymes are proteins. However, certain RNA molecules can be effective biocatalysts too. These RNA molecules have come to be known as ribozymes. In enzymatic reactions, the molecules at the beginning of the process are called substrates, and the enzyme converts them into different molecules, called the products. Almost all processes in a biological cell need enzymes to occur at significant rates. Since enzymes are selective for their substrates and speed up only a few reactions from among many possibilities, the set of enzymes made in a cell determines which metabolic pathways occur in that cell.

Like all catalysts, enzymes work by lowering the activation energy for a reaction, thus dramatically increasing the rate of the reaction. Most enzyme reaction rates are millions of times faster than those of comparable un-catalyzed reactions. As with all catalysts, enzymes are not consumed by the reactions they catalyze, nor do they alter the equilibrium of these reactions. However, enzymes do differ from most other catalysts by being much more specific. Enzymes are known to catalyze about 4,000 biochemical reactions. A few RNA molecules called ribozymes catalyze reactions, with an important example being some parts of the ribosome. Synthetic molecules called artificial enzymes also display enzyme-like catalysis.

Enzyme activity can be affected by other molecules. Inhibitors are molecules that decrease enzyme activity; activators are molecules that increase activity. Many drugs and poisons are enzyme inhibitors. Activity is also affected by temperature, chemical environment (e.g., pH), and the concentration of substrate. Some enzymes are used commercially, for example, in the synthesis of antibiotics. In addition, some household products use enzymes to speed up biochemical reactions.

As early as the late 1700s and early 1800s, the digestion of meat by stomach secretions and the conversion of starch to sugars by plant extracts and saliva were known. However, the mechanism by which this occurred had not been identified. In the 19th century, when studying the fermentation of sugar to alcohol by yeast, Louis Pasteur came to the conclusion that this fermentation was catalyzed by a vital force contained within the yeast cells called "ferments", which were thought to function only within living organisms. He wrote that "alcoholic fermentation is an act correlated with the life and organization of the yeast cells, not with the death or putrefaction of the cells."

In 1878, German physiologist Wilhelm Kühne first used the term enzyme. The word enzyme was used later to refer to nonliving substances such as pepsin, and the word ferment was used to refer to chemical activity produced by living organisms.In 1897, Eduard Buchner began to study the ability of yeast extracts that lacked any living yeast cells to ferment sugar. In a series of experiments at the University of Berlin, he found that the sugar was fermented even when there were no living yeast cells in the mixture. He named the enzyme that brought about the fermentation of sucrose "zymase". In 1907, he received the Nobel Prize in Chemistry "for his biochemical research and his discovery of cell-free fermentation". Following Buchner's example, enzymes are usually named according to the reaction they carry out. Typically, to generate the name of an enzyme, the suffix -ase is added to the name of its substrate (e.g., lactase is the enzyme that cleaves lactose) or the type of reaction (e.g., DNA polymerase forms DNA polymers).

Having shown that enzymes could function outside a living cell, the next step was to determine their biochemical nature. Many early workers noted that enzymatic activity was associated with proteins, but several scientists (such as Nobel laureate Richard Willstätter) argued that proteins were merely carriers for the true enzymes and that proteins per se were incapable of catalysis. However, in 1926, James B. Sumner showed that the enzyme urease was a pure protein and crystallized it; Sumner did likewise for the enzyme catalase in 1937. The conclusion that pure proteins can be enzymes was definitively proved by Northrop and Stanley, who worked on the digestive enzymes pepsin (1930), trypsin and chymotrypsin. These three scientists were awarded the 1946 Nobel Prize in Chemistry.This discovery that enzymes could be crystallized eventually allowed their structures to be solved by x-ray crystallography. This was first done for lysozyme, an enzyme found in tears, saliva and egg whites that digests the coating of some bacteria; the structure was solved by a group led by David Chilton Phillips and published in 1965. This high-resolution structure of lysozyme marked the beginning of the field of structural biology and the effort to understand how enzymes work at an atomic level of detail.

Enzymes are generally globular proteins and range from just 62 amino acid residues in size, for the monomer of 4-oxalocrotonate tautomerase, to over 2,500 residues in the animal fatty acid synthase. A small number of RNA-based biological catalysts exist, with the most common being the ribosome; these are referred to as either RNA-enzymes or ribozymes. The activities of enzymes are determined by their three-dimensional structure. However, although structure does determine function, predicting a novel enzyme's activity just from its structure is a very difficult problem that has not yet been solved.Most enzymes are much larger than the substrates they act on, and only a small portion of the enzyme (around 3-4 amino acids) is directly involved in catalysis. The region that contains these catalytic residues, binds the substrate, and then carries out the reaction is known as the active site. Enzymes can also contain sites that bind cofactors, which are needed for catalysis. Some enzymes also have binding sites for small molecules, which are often direct or indirect products or substrates of the reaction catalyzed. This binding can serve to increase or decrease the enzyme's activity, providing a means for feedback regulation.Like all proteins, enzymes are made as long, linear chains of amino acids that fold to produce a three-dimensional product. Each unique amino acid sequence produces a specific structure, which has unique properties. Individual protein chains may sometimes group together to form a protein complex. Most enzymes can be denatured-that is, unfolded and inactivated-by heating or chemical denaturants, which disrupt the three-dimensional structure of the protein. Depending on the enzyme, denaturation may be reversible or irreversible.

STRUCTURE OF ENZYMES

By the early 1800s, the proteinaceous nature of enzymes had been recognised. Knowledge of the chemistry of proteins drew heavily on the improving techniques and concepts of organic chemistry in the second half of the 1800s; it culminated in the peptide theory of protein structure, usually credited to Fischer and Hofmeister. However, methods that had permitted the separation and synthesis of small peptides were unequal to the task of purifying enzymes. Indeed, there was no consensus that enzymes were proteins. Then, in 1926, Sumner crystallised urease from jack bean meal and announced it to be a simple protein. However, Willstatter argued that enzymes were not proteins but "colloidal carriers" with "active prosthetic groups." However, with the conclusive work by Northrop et al., who isolated a series of crystalline proteolytic enzymes, beginning with pepsin in 1930, the proteinaceous nature of enzymes was established.

The isolation and characterisation of intracellular enzymes was naturally more complicated and, once again, significant improvements were necessary in the separation techniques applicable to proteins before, in the late 1940s, any such enzyme became available in reasonable quantities. Because of the large amounts of accessible starting material and the historical importance of fermentation experiments, most of the first pure intracellular enzymes came from yeast and skeletal muscle. However, as purification methods were improved, the number of enzymes obtained in pure form increased tremendously and still continues to grow. Methods of protein purification are so sophisticated today that, with sufficient effort, any desired enzyme can probably be purified completely, even though very small amounts will be obtained if the source is poor.

Primary Structure

After the protein nature of enzymes had been accepted, the way was clear for more precise analysis of their composition and structure. Most amino acids had been identified by the early 20th century. The methods of amino acid analysis then available, such as gravimetric analysis or microbiological assay, were quite accurate but very slow and required large amounts of material.

The primary structures of pancreatic ribonuclease and egg-white lysozyme were published in 1963. Both of these enzymes, simple extracellular proteins, contain about 120 amino acids. The first intracellular enzyme to have its primary structure determined was glyceraldehyde 3-phosphate dehydrogenase, which has an amino acid sequence of 330 residues and represents a size (250–400 residues) typical of many enzymes. Protein sequencing is increasingly performed by liquid chromatography/mass spectrometry (LC/MS) techniques, and several tools and software packages are now available for protein identification and characterisation. The methods of protein sequence analysis are now so well developed that no real practical deterrent exists, other than time or expense, to determination of the amino acid sequence of any polypeptide chain.

A more recent fundamental concept called proteome (protein complement to a genome) will enable researchers to unravel biochemical and physiological mechanisms of complex multivariate diseases at the functional molecular level. A new discipline, proteomics, complements physical genome research. Proteomics can be defined as "the qualitative and quantitative comparison of proteomes under different conditions to further unravel biological processes".

The fact that enzymes are highly substrate specific and are generally much larger than the substrates on which they act quickly became apparent. The earliest kinetic analyses of enzymatic reactions indicated the formation of transient enzyme–substrate complexes. These observations could be explained easily if the conversion of substrate to product was assumed to occur at a restricted site on an enzyme molecule. This site soon became known as the active center or, as is more common today, the active site.

Particular compounds were found to react with specific amino acid side chains and thus inhibit particular enzymes. This suggested that such side chains might take part in the catalytic mechanisms of these enzymes. An early example was the inhibition of glycolysis or fermentation by iodoacetic acid, which was later recognised as resulting from reaction with a unique cysteine residue of glyceraldehyde 3-phosphate dehydrogenase, which normally carries the substrate in a thioester linkage. Many such group-specific reagents have now been identified as inhibitors of individual enzymes; often they are effective because of the hyper-reactivity of a functionally important side chain in the enzyme's active site. However, a more sophisticated approach to the design of enzyme inhibitors became possible when the reactive group was attached to a substrate; in this way, the specificity of the target enzyme was utilised to achieve selective inhibition of the enzyme. Such active-sitedirected inhibitors have acquired major importance not only academically in the study of enzyme mechanisms but also commercially in the search for a rational approach to selective toxicity or chemotherapy.

Three-Dimensional Structure

Chemical studies showed that the active site of an enzyme consists of a constellation of amino acid side chains brought together spatially from different parts of the polypeptide chain. If this three-dimensional structure was disrupted by denaturation, that is, without breaking any covalent bonds, the biological activity of the enzyme was destroyed. In addition, it was found that all the information required for a protein to fold up spontaneously in solution and reproduce its native shape was contained in its primary structure. This was part of the original "central dogma" of molecular biology.

The X-ray crystallography of proteins demonstrated unequivocally that a given protein has a unique three-dimensional structure. Among the basic design principles was the tendency of hydrophobic amino acid side chains to be associated with the hydrophobic interior of the folded molecule, whereas charged side chains were almost exclusively situated on the hydrophilic exterior or surface. The first high-resolution crystallographic analysis of an enzyme, egg-white lysozyme, confirmed these principles and led to the proposal of a detailed mechanism. The active site was located in a cleft in the structure, which has subsequently proved to be a common feature of active sites. According to this, the enzymatic reaction takes place in a hydrophobic environment, and the successive chemical events involving substrate and protein side chains are not constrained by the ambient conditions of aqueous solution and neutral pH.

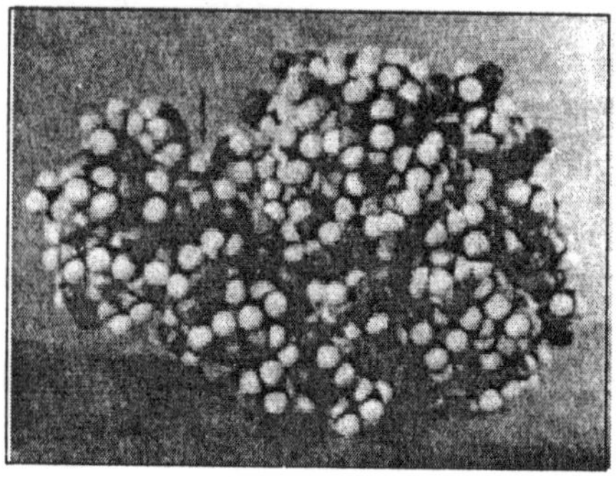

A molecular model of the enzyme lysozyme: the arrow points to the cleft that accepts the polysaccharide substrate

Quaternary structure

Many enzymes consist of more than one polypeptide chain (or subunit), and these must form an aggregate, usually with relatively simple symmetry, before full (or even any) biological activity is conferred. The subunits within an oligomer or multimer are often identical or at least limited to a few different types. Aggregation is generally some form of self-assembly dictated by coherent binding patterns between the subunits, which provide the necessary recognition sites in sorting out the subunits required for assembly.The complexity of this sorting process in a cell becomes evident from the fact that many intracellular enzymes are dimers or tetramers. Increasingly more complicated structures are being recognised and their design principles analysed. These range from enzymes with simple cyclic symmetry up to those with the most elaborate cubic point group symmetry, e.g., octahedral and icosahedral.

The folding of polypeptide chains, along with their aggregation into ordered structures, is a spontaneous process in solution, and this implies that it is exergonic. However, calculation of the time required for a protein to explore all possible structures during the folding process indicates that the search for the "right" structure cannot be entirely random. Thus, even for a small protein such as bovine pancreatic ribonuclease (124 amino acid residues), such a search might take around 1095 years, whereas the experimentally determined time in vivo is a few milliseconds. This dramatic discrepancy led to the concept of kinetic pathways during folding. Such pathways have been experimentally explored, and intermediates identified for various proteins. The stable structure of a protein in solution is therefore identified as the lowest free energy form of the kinetically accessible structures.

A typical enzyme is not an entity completely folded as a whole, as is evident from the growing catalogue of three-dimensional protein structures determined by X-ray crystallography.On the contrary, enzymes frequently consist of apparently autonomous or semiautonomous folding units, called domains. Sometimes, these may be identified as products of limited proteolysis, i.e., regions of the polypeptide chain that can be excised from the chain with retention of their biological properties. Indeed, this has proved in many instances to be a valuable guide to the actual activity contributed by that part of the enzyme. Classical examples of such functional domains can be found in the study of muscle contraction and antibody-antigen recognition. In other cases, domains are not readily released as biologically active entities, and their existence must be inferred from the three-dimensional structure of the enzyme. Most globular proteins can in fact be subdivided into such regions, which generally have molecular masses of 20 000 or less. The active site of an enzyme is often located at the interface between two such domains as, for example, in the well-known cleft of lysozyme or in glutathione reductase. Other domains appear to represent favored folding patterns in the assembly of proteins, but biological activity associated with them can often be inferred from comparison of the structures of related proteins: a typical example is the NAD-binding domain present in dehydrogenases. Structural domains may be regions of the polypeptide chain that fold independently of each other.

CLASSIFICATION OF ENZYMES

Enzymes are divided into six main classes according to the type of reaction catalysed. They are assigned code numbers, prefixed by E.C., which contain four elements separated by points and have the following meaning:

1. the number first indicates to which of the six classes the enzyme belongs,
2. the second indicates the subclass,
3. the third number indicates the sub-subclass, and
4. the fourth is the serial number of the enzyme in its sub-subclass.

The six classes are distinguished in the following manner:

1. *Oxidoreductases:* This class encompasses all enzymes that catalyse redox reactions. The recommended name is dehydrogenase whenever possible, but reductase can also used. Oxidase is used only when O_2 is the acceptor for reduction. The systematic name is formed according to donor : acceptor oxidoreductase.
2. *Transferases:* Transferases catalyse the transfer of a specific group, such as methyl, acyl, amino, glycosyl, or phosphate, fromone substance to another. The recommendedname normally acceptor grouptransferase or donor grouptransferase. The systematic name is formed according to donor : acceptor grouptransferase.
3. *Hydrolases* : Hydrolases catalyse the hydrolytic cleavage of C–O, C–N, C–C, and some other bonds. The recommended name often consists simply of the substrate name with the suffix -ase. The systematic name always includes hydrolase.
4. *Lyases:* Lyases catalyse the cleavage of C–C, C–O, C–N, and other bonds by elimination. The recommended name is, for example, decarboxylase, aldolase, dehydratase elimination of CO_2, aldehyde, and water, respectively). The systematic name formed according to substrate group-lyase.
5. *Isomerases* : Isomerases catalyse geometric or structural rearrangements within a molecule. The different types of isomerism lead to the names racemase, epimerase, isomerase, tautomerase, mutase, or cycloisomerase.
6. *Ligases* : Ligases catalyse the joining of two molecules, coupled with the hydrolysis pyrophosphate bond in ATP or another nucleoside triphosphate. Until 1983, recommended name often included synthetase, but the current recommendation is that names of the type X–Y ligase be used instead, to avoid confusion with the name synthase (which is not confined to enzymes of class 6). The systematic name is formed according to X : Y ligase (ADP-forming).

BIOSYNTHESIS OF ENZYMES

Enzymes are synthesized in cells by the normal machinery of protein synthesis. The structure of any given enzyme is encoded by a structural gene, whose DNA base sequence is transcribed into a messenger RNA, and the mRNA is translated from its triplet code into the amino acid sequence of the desired protein by the ribosomes and associated factors. The enzyme then folds spontaneously into its active conformation.

Posttranslational modifications may be required to target an enzyme to its ultimate intracellular or extracellular location.

For many years, the chemical manipulation of DNAlagged behind that of proteins. The chemical complexity and variety of proteins, with up to 20 different naturally occurring amino acids, served to make them more amenable to increasingly sophisticated methods of analysis. On the other hand, DNA, composed of only four different nucleotides, appeared dauntingly large, with few structural features to make it yield to available methodology. Paradoxically, this very lack of variety in the nature of the constituent nucleotides of DNA has permitted the revolution in genetic engineering, in which the enzymology of DNA has played a prominent part. For example, the discovery and purification of restriction enzymes enabled DNA to be cleaved selectively into defined fragments; phosphatases and ligases permit the fragments to be rejoined selectively; and DNA polymerases allow DNA to be synthesized and sequenced at astonishing speed, all in vitro.

ENZYME KINETICS

The discovery of enzymes is somewhat difficult to determine historically. The discovery is usually attributed to Payen and Persoz, who in 1833 treated an aqueous extract of malt with ethanol and precipitated a heat-labile substance which promoted the hydrolysis of starch. They called their fraction Diastase from the Greek word meaning separation.Much of early enzymology centered on fermentation - an important concern to this day. Many wineries were interested in how to speed up and optimise their fermentation process - to allow for faster product delivery and a greater variety of ferments. There was an early scientific controversy regarding the nature of fermentation in the microorganisms added to juices to induce alcohol production. Pasteur, who was involved in other scientific controversies, regarded fermentation as an essential part of the lifecycle of the microorganism. The total fermentation process - conversion of sugar to alcohol - could not be separated from the living organism.

The chemists of the day, notably Stahl and Liebig, favored a chemical theory of fermentation - that there was in the yeast cell some chemical or group of chemicals which catalysed the conversion of sugar to alcohol. Evidence mounted which supported the chemical theory - in 1860, Berthelow broke open yeast and obtained an alcohol-precipitable fraction which converted sucrose to its two monosaccharide components, glucose and fructose. The suffix "ase" was proposed for use in naming enzymes 5 years later. The controversy between life cycle and chemical arguments came in 1897 when Hans and Edouard Buchner (famous for the Buchner funnel) isolated a cellfree juice which carried out the complete fermentation of sugar. After this little historical background presentation, we will begin a discussion of enzymes and a treatment of enzyme kinetics.

First: A short review of chemical kinetics (is everyone ready for physical chemistry again?

Let's take a simple chemical reaction and discuss how we can describe this reaction both thermodynamically and kinetically:

$$\underset{\textit{substrate}}{A} \rightarrow \underset{\textit{product}}{P}$$

This reaction describes a unireactant system: there is only one reactant, or substrate. If the reaction of interest was A + B → C, this would be a bireactant system.

Let's first describe the kinetics of this system. We want to discuss the rate at which S is converted to P. What is a rate? A reaction rate can be described in several different ways. In this system, the reaction rate can either be described as the formation of product with respect to time (dP/dt) or the disappearance of substrate with respect to time (-dA/dt). In a system such as the one described, these two rate expressions are equivalent. If we were to plot the concentrations of S and P as a function of time from the beginning of the reaction, the plot may look like this.

We can describe the rate of a chemical reaction with a mathematical expression, an equation:

$$2A + B \rightarrow S$$

The rate of this described reaction is a function of the concentrations of all reactants and products. Therefore, a general expression which describes the rate of this chemical reaction would be:

$$-dS/dt = f([A], [B], [S])$$

in its most general sense.

In many cases (that is, in all cases which we'll be interested in), this rate law can be written in the simple form:

$$-dS/dt = k\,[A]^{\alpha}\,[B]^{\beta}\,[S]^{\gamma}$$

k is called the rate constant for the reaction, or the specific rate of the reaction (specific for the case when all concentrations of reactants and products are unity; therefore -dS/dt = k) The coefficients of the concentrations represent the order of the reaction with respect to each concentration. That is, alpha is the order of the reaction with respect to [A], beta is the order of the reaction with respect to [B], etc.

The order of the chemical reaction with respect to each reactant and product simply tells you the relationship between the concentration of the reactants and the rate of the reaction. For example, let's assume that the reaction is first order in A, second order in B, and zeroth order in S. To describe pictorially what this means, if you plot the initial rate of the chemical reaction with respect to one reactant (keeping the concentrations of the other reactants constant), if you plot the reaction rate as a function of A, you will get a linear relationship between [A] and rate. This comes from the mathematics; any quantity to the first power = the quantity. If the quantity is increased some linear amount, the rate will increase some linear amount. If you plot the rate of the reaction relative to the concentration of B (keeping all other concentrations constant), the result is a squared increase in rate resulting from a linear increase in [B].If the reaction is zero order with respect to S, in math, anything to the zeroth power = 1. Therefore, the reaction rate will be independent of the concentration of that quantity; in this case, independent of the concentration of S.

Reaction order is independent of reaction stoichiometry. This 2S term represents reaction stoichiometry; the coefficients represent reaction order.Reaction order is different from the order of a reaction with respect to a reactant. For the example described above, the reaction is first order with respect to A and second order with respect to B. The entire reaction is said to be 3rd order - that is, the reaction order is the sum of the orders of the individual reactants. One example of a first-order reaction is:

$$A \rightarrow P$$

Suppose that the reaction is first order with respect to A, and that no P is present. We are then examining the initial rate of the conversion of S to P. We need not know the order of the reaction with respect to P, because we are not allowing a significant concentration of P to form, and we can ignore it from our rate discussion. This happens to be the set of conditions used to describe reactions in enzyme kinetics. The rate expression for the reaction, based upon these assumptions, is

$$-dS/dt = k\,[S]$$

If you rearrange and integrate this expression, you end up with one of two forms of an integrated rate expression:

$$\ln [S] = -kt + \ln [S_o] \text{ or } [S] = [S_o]\, e^{-kt}$$

For first order reactions, a plot of ln [S] versus *t* will result in a straight line with a slope = -k and a y-intercept = ln $[S_o]$, where $[S_o]$ represents the initial concentration of [S] at the start of the reaction. This reaction primarily as a point of reference to enzymatic reactions. However, there is one simple first-order chemical reaction of importance to both chemists and biochemists, and this is radioactive decay. I'm sure that even some of the biologists in the audience have worked with radioisotopes.

When a substance is radioactive, this means that it has the ability to lose energy in the form of radiative energy to decay to some nonstable nonradioactive isotope. This radioactive decay is a first order process, dependent upon the concentration of the radioactive species. One way we characterise radioisotopes is by their half lives. The half life of a radioisotope is the amount of time necessary for half of the radioactive molecules to decay to some stable isotope. In math terms, when half of the initial molecules have decayed, then $[S]/[S_o] = 0.5$, and

$$0.5 = e^{-kt} \ln (0.5) = -kt_{1/2} \; t_{1/2} = 0.693/k$$

Reaction Equilibria

Let's look at another simple reaction:

$$A\text{-}B + C \rightarrow A + B\text{-}C$$

This time, let's examining the thermodynamics of the reaction. Why does this reaction take place? The conversion of A-B to B-C is governed thermodynamically by the difference in free energy between reactants and products. We can draw an energy diagram to describe the reaction, and we'll discuss three different reaction scenarios.

1. The free energy associated with B-C is much less than the free energy associated with A-B. In this case, the change in free energy (ΔG) between product and reactant is highly negative. At equilibrium, the free energy difference between reactants and products is given by the expression:

   $\Delta G_o = -RT \ln K$, where ΔG_o is the free energy difference at equilibrium, R is the gas constant, T is the absolute temperature, and K is the equilibrium constant for the reaction. If the free energy difference is large, then the equilibrium constant for the reaction is large, and the concentration of B-C will be at equilibrium much greater than the concentration of A-B. That is, the reaction will proceed until most of the A-B is converted to B-C.

2. The free energy associated with B-C is slightly less than that associated with AB.

   Therefore, there is little free energy difference between reactants and products. At equilibrium, there will only be a little more B-C than A-B.

3. The free energy associated with B-C is slightly greater than that associated with A-B. Since we are starting with A-B, the reaction will still proceed as written until the equilibrium concentrations are reached.

4. The free energy associated with B-C is much greater than A-B. Here the reaction is thermodynamically unfavorable, and will not proceed appreciably. This thermodynamic argument describes the equilibrium concentrations of A-B and B-C beginning with A-B. The free energy difference is related to these concentrations.

What's the relationship between equilibria and rate? There is not necessarily any relationship between reaction equilibrium and rate, as we will see in a few minutes.

Enzymatic Reactions

An enzyme reaction is exactly like a chemical reaction. Now, the fact that a chemical reaction has a negative .Go does not mean that it will progress at a measurable rate. The rate constant can not be determined by the sign or magnitude of ΔG^o.

$$\text{Example: glucose } (C_6H_{12}) + 6\ O_2 \rightarrow 6\ CO_2 + 6\ H_2O$$

$$\Delta G^{o\prime} = -686 \text{ kcal/mol glucose}$$

This reaction is extremely favorable thermodynamically. However, if you leave glucose out in a bowl at room temperature, where it's exposed to lots of oxygen, the only thing you get are ants. Why doesn't glucose spontaneously convert to products? There must be some energy barrier between reactants and products which prohibits the conversion of glucose to CO_2 and H_2O. This energy barrier is called the activation energy for the reaction. Another way to express this phenomenon - the reaction is thermodynamically favorable (negative $\Delta G^{o\prime}$) but kinetically unfavorable (does not occur at appreciable rate due to energy barrier). Here is an overhead describing the energetics of the problem. There are two ways to overcome this kinetic barrier:

1. Add energy to the system. This increases the energy of the reactants such that the energy barrier which the reactants must overcome is smaller. If you set a match to glucose, it will burn, resulting in the proper products.
2. Add a catalyst.

In biological systems, an enzyme is nothing more than a catalyst - it speeds up the rate of a chemical reaction without itself being consumed. How does an enzyme act as a catalyst? - It reduces the activation energy for a chemical reaction.The chemical equilibrium for the reaction is not altered. The final concentration 'of reactants and products is determined solely by the equilibrium constant for the reaction, which is proportional to the free energy difference between reactants and products. The enzyme hastens the equilibration of the system by lowering the potential energy barrier between reactants and products.

Enzyme Active Sites

The enzyme-driven conversion of substrate to products can be written as:

$$S \rightarrow P$$

where E is a nonconsumed catalyst. However, somehow during the chemical reaction the substrate needs to combine with the catalyst, the enzyme. Therefore the total reaction might be written:

$$E + S \rightarrow ES \rightarrow E + P$$

where ES is an enzyme-substrate complex.

The high degree of substrate specificity in enzymes has led to speculation regarding the nature of the substrate binding site on proteins. In 1894, Emil Fischer suggested the template or lock and key analogy of substrate-enzyme interactions. This theory assumes that the enzyme possesses some region which is complementary in size, shape, and chemical nature to its substrate. This region of the enzyme is called the substrate binding site, catalytic site, or substrate binding site. Because of the exact nature of the substrate binding site on proteins, only a single substrate or single small range of substrates can bind to an enzyme and act as a substrate. A second model for substrate binding was proposed by Koshland -called the induced fit hypothesis. Substrate binding causes a modification in enzyme conformation such that the enzyme binds the substrate more tightly, or that the change in enzyme conformation aligns the bound substrate to other substrates or protein functional groups.This, by the way, is the same Koshland who developed the sequential model for allosteric proteins. The induced fit model says that there is a change in conformation of a protein upon ligand binding - which was one of the assumptions upon which the sequential model for allosteric proteins was based. Here is a different representation of the lock and key model. The substrate may have some chemical features such as an amine group and a hydrophobic substituent. The enzyme would then have features that are complementary to the substrate. For example, there may be a carbonyl group or hydroxyl group on the enzyme that may hydrogen bond with the amine group on the substrate. There may also be a hydrophobic pocket in the enzyme that would form Van der Waals type interactions with the substrate.

Kinetics of Unireactant Systems

Henri presented the first general rate equation for reactions involving enzymes. The derivation of Henri's equation arose from the following assumptions:

1. The enzyme is a catalyst.
2. Enzyme and substrate react rapidly to form an enzyme-substrate complex.
3. Only a single substrate and single enzyme-substrate complex are involved and that the complex breaks down directly to form free enzyme and product.
4. Enzyme, substrate, and enzyme-substrate complex are at equilibrium.
5. Concentration of substrate is much larger than concentration of enzyme so that the formation of the enzyme-substrate complex does not significantly deplete the concentration of substrate.
6. The overall rate of the reaction is limited by the breakdown of the enzymesubstrate complex to form free enzyme and product.

7. The velocity is measured during the early stages of the reaction so that the reverse reaction is insignificant.

The major assumption in this method is all of the components of the reaction system are at equilibrium - this is called the quasi-equilibrium or rapid equilibrium assumption. The reaction can be depicted as:

$$E + S \_ ES \rightarrow E + P$$

and the derived Henri equation is:

$$v = K[S]/(1 + [S]/K_S)$$

where [S] = fixed substrate concentration; v = initial velocity (d[P]/dt or -d[S]/dt) - in practice, v approximately equal to $\Delta[P]/\Delta t$ or $-\Delta[S]/\Delta t$, provided the appearance of P is linear with time for the duration of the assay and no more than 5% of the original S is utilised.

$$K_S = \text{dissociation constant of ES} = k_{-1}/k_1 = [E][S]/[ES]$$

K = constant characteristic of the particular enzyme preparation.

Ten years later, Michaelis and Menton presented a slightly modified version of the rate equation:

$$v = k_2[E]_t[S]/K_S + [S]$$

If $v = k_2[ES]$, then $k_2[E]_t = V_{max}$, or, the limiting maximal velocity is observed when all the enzyme exists as ES. Substituting,

$$v = V_{max}[S]/(K_S + [S]) \text{ or } v/V_{max} = [S]/(K_S + [S])$$

Derivation of this expression:

$$E + S \_ ES \rightarrow E + P \quad (1)$$

$$\text{Mass balance eq. - } [E]t = [E] + [ES] \quad (2)$$

Velocity dependence - $v = k_{cat}[ES]$ (note that k_{cat} is the rate constant describing the rate limiting step of the enzyme catalysed reaction). The Henri-Michaelis-Menton equation derived above requires rapid equilibrium between S, E, and ES. Briggs and Haldane later rederived the same relationship without depending upon the rapid equilibrium assumption. The Briggs-Haldane approach:

$$E + S \rightarrow ES$$

ES may decompose by one of two processes:

$$ES \rightarrow E + S$$

$$ES \rightarrow E + P$$

Therefore, the rate of ES formation = and the rate of ES decomposition =

The Briggs-Haldane expression works because of the steady-state assumption. At the steady state, the rate of appearance of ES is equal to the rate of its disappearance, or

$$+d[ES]/dt = -d[ES]/dt$$

Going through the derivation, you get

$$v = V_{max}[S]/(K_M + [S])$$

K_M is a dynamic or pseudoequilibrium constant

$$K_M = k_{-1} + k_{cat}/ k_1$$

; K_S is a true equilibrium constant.

The Briggs-Haldane equation is equivalent to the Henri-Michaelis-Menton equation except for this constant. The equilibrium constant is related to K_M when $k_{cat} << k_{-1}$. Then,

$$K_M = k_{-1} + k_{cat}/ k_1 = K_M = k_{-1} / k_1 = K_S$$

An enzyme that obeys Michaelis-Menton kinetics will give a hyperbolic relationship between velocity and substrate concentration. One relationship from both Michaelis-Menton and Briggs-Haldane treatments of enzyme kinetics is the situation where K_M (or K_S) = [S]. At this point, then v = ([S]/2[S])V_{max} = ½(V_{max}). Therefore, if you have a curve of velocity versus substrate concentration, you could estimate KM from this curve. Why is the knowledge of K_M important for the study of enzymes?

1. The knowledge of K_M establishes an approximate value for the intracellular substrate concentration under normal conditions. Enzymes are normally designed to work with their K_M's about the same as the intracellular substrate concentration. This represents a compromise for the best use of the enzyme - that is, some of the sensitivity of the enzyme to substrate concentration is conserved without entire loss of the catalytic efficiency of the enzyme.

- If $[S] << K_M$, then the enzyme rate will be very sensitive to the substrate concentration, but the catalytic efficiency of the enzyme will be wasted.
- If $[S] >> K_M$, then the enzyme will quickly convert substrate to product, but will be relatively insensitive to substrate concentration (and therefore will be more difficult to regulate).

2. The K_M of an enzyme provides some mean to compare different enzymes. For example, you might want to compare the same enzyme isolated from different organisms, or the same enzyme isolated from different tissues in the same organism. K_M is one measurement to make for the sake of comparison.
3. A ligand-induced change in K_M is one way to affect enzyme activity. We'll discuss this when we begin to talk about inhibition of enzymes.

Reaction order of Enzymatic Reactions

Let's go back and reexamine the shape of the velocity versus substrate curve for enzymatic reactions, based upon what we have covered on ligand binding, catalytic efficiency, and reaction order. First let's just superficially look at this curve and see what we can derive from it regarding reaction order. At low substrate concentrations, it appears that a linear increase in substrate concentration results in a linear increase in reaction rate. If you return to the Michaelis-Menton expression:

$$v = k_{cat}[E]_t[S]/(K_M + [S])$$

you see that at low substrate concentrations $[S] << K_M$, and therefore $([S] + K_M)$ is approximately equal to K_M. Under these conditions, the reaction is approximately first order with respect to S. At high substrate concentrations, there appears to be little change in reaction rate in response to addition of substrate. At high substrate concentrations, $([S] + K_M)$ is approximately equal to [S], and the rate expression simplifies to:

$$v = k_{cat}[E]_t/K_M$$

Here, there is no effect of changes in [S] on reaction rate.

Graphical Determination of V_{max} and K_M

The v versus [S] curve is a hyperbola - it is difficult to determine V_{max} and K_M from this graph. The first useful way to determine these two constants graphically was proposed by Lineweaver and Burk in 1934 - called the Lineweaver Burk plot (appropriately enough) or the double reciprocal plot. This plot gives a linear representation of v versus [S] data, allowing for the simple determination of V_{max} and K_M. The double reciprocal arrangement is a simple rearrangement of the Michaelis-Menton equation:

$$1/v = (K_M/V_{max})(1/[S]) + 1/V_{max}.$$

This equation is now in a simple y = mx + b form, allowing simple determination of the slope and intercept.

Enzyme Inhibition

There are three types of simple inhibition of unireactant enzyme systems, which are:

- Competitive inhibition
- Noncompetitive inhibition
- Uncompetitive inhibition

Simple Competitive Inhibition

A competitive inhibitor of an enzyme is a substance which combines with the enzyme in such a way to prevent binding of normal substrate. A competitive inhibitor might be

- a nonmetabolisable analog or derivative of the substrate, - an alternative substrate, or
- a product.

Your textbook uses as an example the inhibition of succinate dehydrogenase - catalyzes the dehydrogenation of succinate to fumerate. Malonic acid is a competitive inhibitor. As you can see from its structure, malonic acid is somewhat similar to succinic acid, except that it is missing one methylene group. Because of this, there is no site for double bond formation - the material binds to the enzyme, but is not converted to product. Here is a second example involving the enzyme HMG-CoA reductase. This enzyme catalyses the committed step in cholesterol biosynthesis in the liver. The multibillion dollar drug Lipitor looks like the normal substrate, HMG-CoA, except that the group that is reduced (the thioester in HMG-CoA) is only an alcohol in the inhibitor. The inhibitor looks enough like the substrate to bind to the active site, but cannot be converted into a product.

Alternative Substrates can also inhibit a measured enzyme reaction. Example - Hexokinase - three acceptable substrates for hexokinase are glucose, fructose, and mannose (all 6-carbon sugars). In assaying hexokinase, you might measure the formation of a single produce, for example, glucose-6-phosphate. If the glucose you use to assay the enzyme is contaminated with fructose, the measured activity will be lower than in the presence of pure glucose - apparent enzyme inhibition.You can describe competitive inhibition by drawing appropriate equilibrium expressions:

$$EI \leftrightarrow E + S \leftrightarrow ES \leftrightarrow EP \leftrightarrow E + P$$

This is still a unireactant enzyme we're describing. [EI] represents the concentration of the enzyme-inhibitor complex - this complex, in competitive inhibition, cannot bind substrate.

K_i is the dissociation constant for inhibitor binding to enzyme.

$$K_i = [E][I]/[EI] \quad K_S = [E][S]/[ES]$$

If we assume rapid equilibrium between E, S, I, ES, and EI:

1. v_i - the measured velocity of the enzyme in the presence of the inhibitor I - can be made equal to the velocity observed in the absence of I simply by adding more substrate [S].
2. Regardless of inhibitor concentration, the enzyme can be driven entirely into the ES form simply by adding infinite substrate.

Result - The presence of an inhibitor does not change the V_{max} for the enzyme However, in the presence of I, there will be some enzyme in EI - this cannot bind substrate. The result is that the measured K_M in the presence of I, the K_M apparent, is greater in the presence of inhibitor than in its absence. We can derive a rate expression relating v, V_{max}, [S], KM, I, and KI using either the rapid equilibrium or steady state assumption:

$$v = V_{max}[S]/\{K_s(1 + [I]/K_i) + [S]\}$$

Double reciprocal equation:

$$1/v = \{K_s(1 + [I]/K_i)/ V_{max}[S]\} + [S]/ V_{max}[S]$$
$$1/v = (K_s/ V_{max}) (1 + [I]/K_i) (1/[S])$$

This shows mathematically what I tried to explain before - in the presence of a competitive inhibitor, there is no affect on V_{max}. However, the expression describing substrate binding, K_s is altered by this term $(1 + [I]/K_i)$.

This means that the binding affinity of the enzyme for substrate is less in the presence of inhibitor than in its absence, or $K_{M\ apparent} > K_M$. If the steady state assumption is used rather than the rapid equilibrium assumption, the derived relationship is the same, except K_S will be replaced by K_M. What affect does the presence of a competitive inhibitor have upon an enzyme catalysed reaction? Assume that I is added to a reaction mixture to [I] = $3K_I$ .Therefore

$$K_{M\ apparent} = K_M(1 + 3/1) = 4K_M.$$

Therefore, in the presence of the inhibitor it takes 4 times as much substrate to attain 0.5 V_{max}.

Graphical Analysis of Competitive Inhibition Systems

Reciprocal rate expression for an enzyme in the presence of [I] is shown on the handout. From examination of above, the slope of the reciprocal plot in the presence of inhibitor is greater than in its absence by the factor $(1 + [I]/K_I)$, but the yintercept remains the same. You can plot the reciprocal of enzyme velocity as a function of reciprocal substrate concentration at any given [I], and the result will be a series of straight lines connected at the y-intercept. From the slope of any reciprocal plot in the presence of inhibitor compared to the slope in the absence of inhibitor, K_I can be determined.

Noncompetitive Inhibition

A noncompetitive inhibitor of an enzyme has no effect upon substrate binding, and the substrate has no effect upon inhibitor binding. The substrate and inhibitor bind randomly, reversibly, and independently at different sites. In other words:

I binds to both E and ES

S binds to both E and EI

There are four possible enzyme species, all at equilibrium, in noncompetitive inhibition. The equilibria are shown on the handout. This shows, by inspection of the equilibrium complexes, that the binding of inhibitor does not affect the binding of substrate, and vice versa. That is, the dissociation constant Ki for the inhibitor is the same for dissociation of I from E and from ES. The same holds for substrate - K_S is the same whether or not dissociation from E or ESI is described. Also, from examination of these equilibrium expressions it becomes apparent that the addition of an infinite amount of substrate will not drive the enzyme entirely into the ES form. At infinite substrate, the enzyme will be at equilibrium between ES and ESI, and that the relative amounts of these two species will depend entirely on the so-called specific inhibitor concentration, that is, the concentration of the inhibitor relative to its dissociation constant K_i. Therefore, V_{max} in the presence of I will be less than V_{max} in the absence of I. You can get some type of pictorial representation of noncompetitive inhibition from the drawings on today's handout. I does not hinder the binding of S in this representation; the only thing that I does is that when bound to the enzyme I prohibits the enzyme from converting S to P.

We can derive a rate expression which relates v, V_{max}, [S], [I], KS, and Ki using the rapid equilibrium assumption. The equation is provided on the handout. Examination of this expression shows that in the denominator both the KS and the [S] terms are reduced by the factor $(1 + [I]/K_i)$. We can get a better picture of how a noncompetitive inhibitor affects the rate expression for an enzyme by dividing both denominators by the factor $(1+[I]/K_i)$.As we deduced from looking at the equilibria, the affect which a noncompetitive inhibitor has upon an enzyme is to decrease its apparent V_{max}. It's clear from this expression that adding large concentrations of S has no affect upon the maximal velocity, only the ratio of I to K_i. The derivation of this expression for noncompetitive inhibition can only be done using the rapid equilibrium assumption. The derivation of a rate expression for noncompetitive inhibition using the steady state assumption results in a complex expression which includes $[I]^2$ and $[S]^2$ terms.

Reciprocal Plot for Noncompetitive Inhibition

The reciprocal form of the rate expression describing noncompetitive inhibition is shown on the handout. This expression indicates that both the slope and the 1/v intercept of a 1/v versus 1/[S] plot is increased by the factor (1 + $[I]/K_i$) in the presence of a noncompetitive inhibitor. If the slope and the yintercept of a plot increase by the same factor, then the 1/[S] intercept of the plot should remain the same (as you might intuitively think that it should, since the 1/[S] intercept is equal to $-1/K_M$, and we've already said that the K_M does not change in the presence of a noncompetitive inhibitor).

Mixed Inhibition

Noncompetitive inhibition is a specialised form of mixed inhibition. For noncompetitive inhibition, we state that the binding of the substrate to the enzyme has absolutely no effect on the binding of inhibitor to the enzyme, and vice versa. In truth, this seldom occurs. There is usually a little effect of the inhibitor on enzyme binding and vice versa, which means that there are two Ki's and two Ks's. The result is that the math gets much more complicated. Graphically, mixed inhibition is fairly easily interpreted. If the lines generated in the presence of several inhibitor concentrations intersect on the y axis, you have competitive inhibition. If the lines intercept on the x axis, you have noncompetitive inhibition. If the lines intercept somewhere other than on the axes, you have mixed inhibition.

Uncompetitive Inhibition

In the case of uncompetitive inhibition, the inhibitor binds only to the ES complex, and not to free E. This type of inhibition is rare in unireactant systems, but worth considering at this time because it is a simple example of the addition of two enzyme ligands in obligate order.In multireactant systems, uncompetitive inhibition is common - the inhibitor will be uncompetitive if it binds to the enzyme after binding of substrate. In multireactant systems, this type of binding is more common because the inhibitor has a greater number of choices of enzyme species which to bind to. Classic uncompetitive inhibition is characterised by the equilibrium shown on the handout. At any [I], infinite substrate will not drive all of the enzyme to ES. Therefore, there will always be some nonproductive complex present. Therefore, V_{max} is affected by the presence of uncompetitive [I].

Unlike noncompetitive inhibition, [I] affects K_M - the addition of [I] will drive the equilibrium E + S. ES to the right by removing ES. The apparent result will be an increase in the affinity of the enzyme for substrate, or an apparent decrease in K_M.You can (as we have consistently) derive an expression relating v, V_{max}, [S], [I], K_S, and K_i:

$$v = k_2[ES]\ [E]_t = [E] + [ES] + [ESI].$$

The final expression is shown on the handout. This expression does not contain an [I]/K_i term because no EI complex is formed. An uncompetitive inhibitor decreases both V_{max} and K_M to the same extent. The reciprocal plot can be done with uncompetitive inhibition. The necessary relationship is shown on the handout. The slope of the reciprocal plot is not changed in the presence of an uncompetitive inhibitor, since both the V_{max} and the K_M are changed in the same amount - the changes cancel. The y-axis intercept changes, since it is described solely by $1/V_{max}$ app. The result is a series of parallel lines whose y-axis intercept is dependent upon the concentration of [I].

Environmental Effectors of Enzyme Activity

1. *Temperature*—An enzymatic reaction can also be sped up by increasing the temperature, but only to a defined limit. A graph of temperature versus reaction rate for a chemical reaction will steadily increase as a function of temperature. A enzymatic reaction rate will increase to a point, and then precipitously fall to zero. The temperature at which the reaction ceases is the temperature at which the enzyme denatures. In its unfolded form, the substrate binding site and catalytic activity of the enzyme no longer exist.
2. *pH*—The effect of pH on activity for any particular enzyme depends on the pK of the ionizing groups on that enzyme, and on ionising groups of the substrate in the enzymatic reaction. The pH optimum for the enzyme is usually a function of the enzyme rather than the substrate. For example, here are the pH profiles of three endopeptidases we discussed when we talked about protein sequencing. The substrate for these enzymes, that is, proteins, is essentially the same; however the pH profile of the enzyme activity reflects the nature of the specific enzyme and its environment.

The pH sensitivity of an enzyme is usually caused by a change in the protonation state of an amino acid sidechain at the enzyme active site. However, the rate of an enzymatic reaction may also be altered due to changes in the three-dimensional structure of the enzyme resulting from protonation distant from the active site.

Multisubstrate Reactions

Suppose that we have a reaction where two substrates are converted into two products. There are several different ways in which these substrates can bind to the enzyme.

- *Random:* In this case, it does not matter which substrate binds to the enzyme first. An example of this is the enzyme hexokinase, which transfers a phosphate group from ATP to glucose to yield ADP and glucose-6-phosphate.
- *Ordered:* In this case, one substrate must bind before another. You can imagine a long binding site in an enzyme where one substrate lies in the bottom of the trough while

the second binds on top of it, preventing the first from leaving. An example of an enzyme that works by this mechanism is alcohol dehydrogenase, where ethanol binds first to the enzyme and the NAD molecule binds second.

– *Ping Pong:* Here, one substrate binds, is reacted, and the product released before the second substrate binds. A classical example is the enzyme nucleoside diphosphokinase, which catalyses the transfer of a phosphate group from ATP to another nucleoside diphosphate such as GDP:

$$ATP + E \rightarrow ADP + E\text{-}P_i$$

$$GDP + E\text{-}P_i \rightarrow GTP + E$$

Catalysts increase the rate of otherwise slow or imperceptible reactions without undergoing any net change in their structure. The early development of the concept of catalysis in the 19th century went hand in hand with the discovery of powerful catalysts from biological sources. These were called enzymes and were later found to be proteins. They mediate all synthetic and degradative reactions carried out by living organisms. They are very efficient catalysts, often far superior to conventional chemical catalysts, for which reason they are being employed increasingly in today's high-technological society, as a highly significant part of biotechnological expansion. Their utilization has created a billion dollar business including a wide diversity of industrial processes, consumer products, and the burgeoning field of biosensors. Further applications are being discovered constantly.

Enzymes have a number of distinct advantages over conventional chemical catalysts. Foremost amongst these are their specificity and selectivity not only for particular reactions but also in their discrimination between similar parts of molecules (regiospecificity) or optical isomers (stereospecificity). They catalyse only the reactions of very narrow ranges of reactants (substrates), which may consist of a small number of closely related classes of compounds (e.g. trypsin catalyses the hydrolysis of some peptides and esters in addition to most proteins), a single class of compounds (e.g. hexokinase catalyses the transfer of a phosphate group from ATP to several hexoses), or a single compound (e.g. glucose oxidase oxidises only glucose amongst the naturally occurring sugars). This means that the chosen reaction can be catalysed to the exclusion of side-reactions, eliminating undesirable by-products. Thus, higher productivities may be achieved, reducing material costs. As a bonus, the product is generated in an uncontaminated state so reducing purification costs and the downstream environmental burden. Often a smaller number of steps may be required to produce the desired end-product. In addition, certain stereospecific reactions cannot be achieved by classical chemical methods without a large expenditure of time and effort.

Enzymes work under generally mild processing conditions of temperature, pressure and pH. This decreases the energy requirements, reduces the capital costs due to corrosion-resistant process equipment and further reduces unwanted side-reactions. The high reaction velocities and straightforward catalytic regulation achieved in enzyme-catalysed reactions allow an increase in productivity with reduced manufacturing costs due to wages and overheads. There are some disadvantages in the use of enzymes which cannot be ignored but which are currently being addressed and overcome. In particular, the high cost of enzyme isolation and purification still discourages their use, especially in areas which currently have an established alternative procedure. The generally unstable nature of enzymes, when removed from their natural environment, is also a major drawback to their more extensive use.

The amount of enzyme present or used in a process is difficult to determine in absolute terms (e.g. grams), as its purity is often low and a proportion may be in an inactive, or partially active, state. More relevant parameters are the activity of the enzyme preparation and the activities of any contaminating enzymes. These activities are usually measured in terms of the activity unit (U) which is defined as the amount which will catalyse the transformation of 1 micromole of the substrate per minute under standard conditions. Typically, this represents 10^{-6} - 10^{-11} Kg for pure enzymes and 10^{-4} - 10^{-7} Kg for industrial enzyme preparations. Another unit of enzyme activity has been recommended. This is the katal (kat) which is defined as the amount which will catalyse the transformation of one mole of substance per second (1 kat = 60 000 000 U). It is an impracticable unit and has not yet received widespread acceptance.

Sometimes non-standard activity units are used, such as Soxhet, Anson and Kilo Novo units, which are based on physical changes such as lowering viscosity and supposedly better understood by industry. Rightfully, such units are gradually falling into disuse. The activity is a measure of enzyme content that is clearly of major interest when the enzyme is to be used in a process. For this reason, enzymes are usually marketed in terms of activity rather than weight. The specific activity (e.g. U Kg^{-1}) is a parameter of interest, some utility as an index of purity but lesser importance.

There is a major problem with these definitions of activity; the rather vague notion of "standard conditions". These are meant to refer to optimal conditions, especially with regard to pH, ionic strength, temperature, substrate concentration and the presence and concentration of cofactors and coenzymes. However, these so-termed optimal conditions vary both between laboratories and between suppliers. They also depend on the particular application in which the enzyme is to be used. Additionally, preparations of the same notional specific activity may differ with respect to stability and be capable of very different total catalytic productivity (this is the total substrate converted to product during the lifetime of the catalyst, under specified conditions). Conditions for maximum initial activity are not necessarily those for maximum stability. Great care has to be taken over the consideration of these factors when the most efficient catalyst for a particular purpose is to be chosen.

There are many directions in which enzyme technologists are currently applying their art and which are at the forefront of biotechnological research and development. At present, relatively few enzymes are available on a large scale (i.e. > kg) and are suitable for industrial applications. These shortcomings are being addressed in a number of ways:

1. new enzymes are being sought in the natural environment and by strain selection;
2. established industrial enzymes are being used in as wide a variety of ways as can be conceived;
3. novel enzymes are being designed and produce by genetic engineering;
4. new organic catalysts are being designed and synthesised using the 'knowhow' established from enzymology; and
5. more complex enzyme systems are being utilised.

Each of these areas has a extensive and rapidly expanding literature. Some advances possibly belong more properly to other areas of science. Thus, the development of genetically improved enzymes is generally undertaken by molecular biologists and the design and synthesis of novel enzyme-like catalysts is in the provenance of the organic chemists. Both groups of workers will, however, base their science on data provided by the enzyme technologist.

Many enzymes are not totally specific for their natural substrates. Some have been found to catalyse reactions quite different from those given as the normal reactions and reflected in their name and EC number. Sometimes it necessary to place the enzyme in an unusual environment in order to display new activities. Thus, lipases act as transesterases in primarily non-aqueous environments. In other cases, changes in the environment are not necessary.

Glucose oxidase is specific for its reducing substrate (D-glucose) but fairly non-specific in its choice of oxidant, normally molecular oxygen. It has been established that benzoquinone is also an effective electron-accepting substrate:

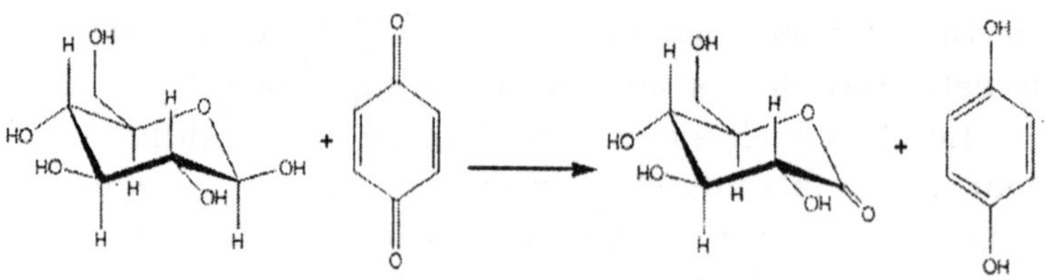

[1]

b-D-glucose + benzoquinone®D-glucono-1,5-lactone + hydroquinone

The product of this reaction, hydroquinone, is a valuable organic chemical, being used in the photographic industry and as an antioxidant. The reaction gives nearly 100% yields, with no possibility of the peroxide-induced inactivation which occurs using molecular oxygen as oxidant. Because of the ready solubility of benzoquinone and low solubility of molecular oxygen, the above reaction ([1]) can give productivity rates several times greater than the 'natural' reaction.Acetylcholinesterase normally catalyses the hydrolysis of acetylcholine, the excitatory neurotransmitter, in the synaptic junctions of vertebrates.

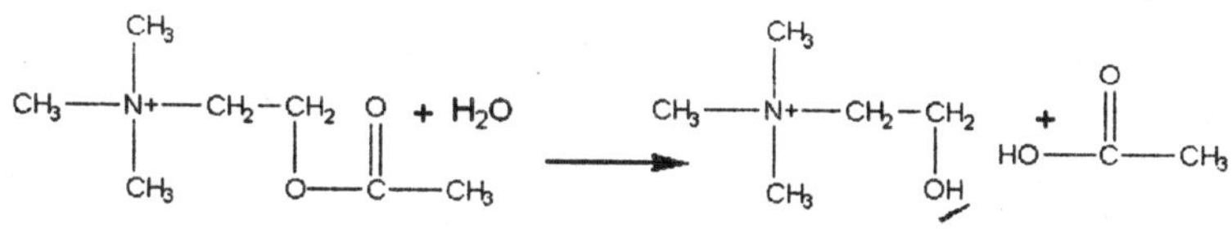

[2]

acetylcholine + water→ choline + acetic acid

The acetylcholinesterase from the electric eel has been found additionally to catalyse the stereospecific hydrolysis of acetyl-D-carnitine but not acetyl-L-carnitine.

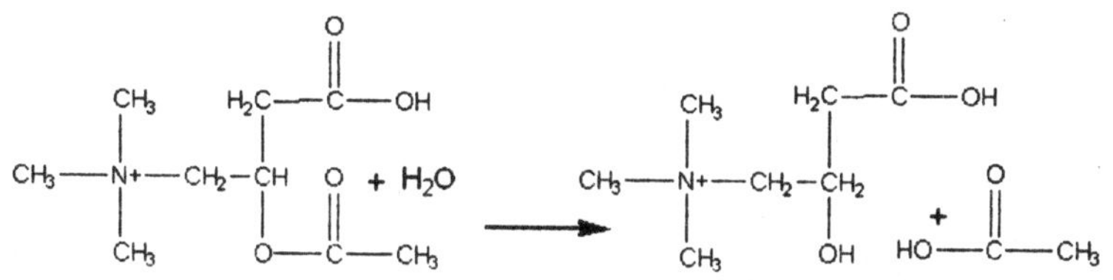

[3]

acetyl-D-carnitine + water→ D-Carnitine + acetic acid

Although the reaction involving acetyl-D-carnitine has a second-order rate (specificity) constant four orders of magnitude smaller than that utilising acetylcholine, the productivities at high substrate concentrations are comparable, the 'natural' substrate, acetylcholine, causing pronounced substrate inhibition which is not apparent with acetyl-D-carnitine. The 'unnatural' reaction is useful as it may be used in the preparation of L-carnitine, one of the vitamins that has numerous therapeutic applications; the D-isomer being biologically inactive. The enzyme can, therefore, be used in a manner similar to the way in which aminoacylase is used to resolve racemic amino acids.

Chemically synthesised racemic DL-carnitine may be acetylated, using acetyl chloride, to give acetyl-DL-carnitine. The acetylcholinesterase may then be used to produce a mixture of acetyl-L-carnitine and D-carnitine that may be simply resolved by ion-exchange chromatography. The acetyl-L-carnitine is as biologically active as L-carnitine and may be used directly.

Enzyme engineering (or enzyme technology) is the application of fermentation process using catalytic activity of isolated enzymes, to produce new metabolites or to convert some compounds into another's (biotransformation). These products will be useful as chemicals, pharmaceutiacls, fuel, food or agricultural additives. Enzyme reactor consists of a vessels, containing a reactional medium, used to perform a desired conversion by enzymatic means. Enzymes used in this process are free in the solution or immobilised in particulate, membranous or fibrous support.

There are a number of properties which may be improved or altered by genetic engineering including the yield and kinetics of the enzyme, the ease of downstream processing and various safety aspects. Enzymes from dangerous or unapproved microorganisms and from slow growing or limited plant or animal tissue may be cloned into safe high-production microorganisms. In the future, enzymes may be redesigned to fit more appropriately into industrial processes; for example, making glucose isomerase less susceptible to inhibition by the Ca^{2+} present in the starch saccharification processing stream.

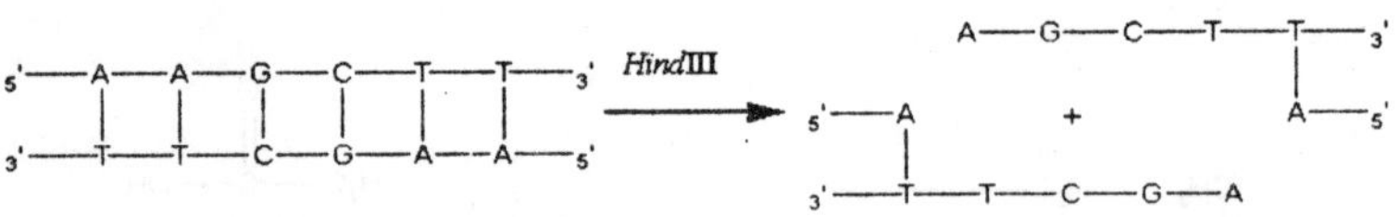

[4]

Intact DNA → cleaved DNA

The amount of enzyme produced by a microorganism may be increased by increasing the number of gene copies that code for it. This principle has been used to increase the activity of penicillin-G-amidase in Escherichia coli. The cellular DNA from a producing strain is selectively cleaved by the restriction endonuclease HindIII. This hydrolyses the DNA at relatively rare sites containing the 5'-AAGCTT-3' base sequence to give identical 'staggered' ends. The total DNA is cleaved into about 10000 fragments, only one of which contains the required genetic information. These fragments are individual cloned into a cosmid vector and thereby returned to E. coli. These colonies containing the active gene are identified by their inhibition of a 6-amino-penicillanic acid-sensitive organism. Such colonies are isolated and the penicillin-G-amidase gene transferred on to pBR322 plasmids and recloned back into E. coli.

The engineered cells, aided by the plasmid amplification at around 50 copies per cell, produce penicillin-G-amidase constitutively and in considerably higher quantities than does the fully induced parental strain. Such increased yields are economically relevant not just for the increased volumetric productivity but also because of reduced downstream processing costs, the resulting crude enzyme being that much purer.

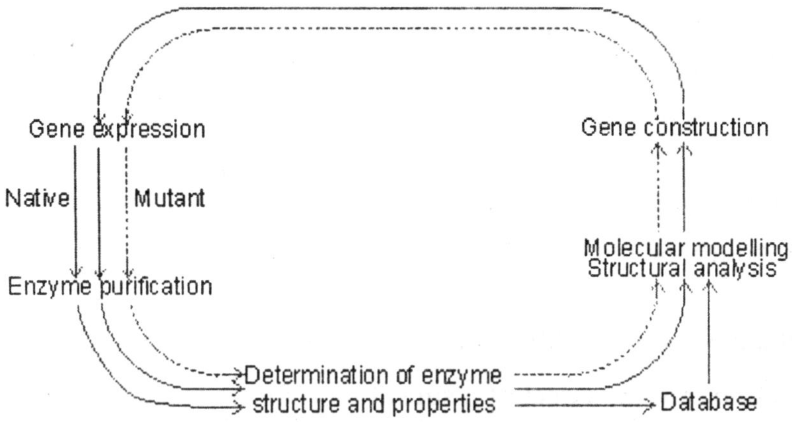

The protein engineering cycle

The process starts with the isolation and characterisation of the required enzyme. This information is analysed together with the database of known and putative structural effects of amino acid substitutions to produce a possible improved structure. This factitious enzyme is constructed by site-directed mutagenesis, isolated and characterised. The results, successful or unsuccessful, are added to the database, and the process repeated until the required result is obtained. Another extremely promising area of genetic engineering is protein engineering. New enzyme structures may be designed and produced in order to improve on existing enzymes or create new activities.

As indicated by the method used for site-directed mutagenesis, the preferred pathway for creating new enzymes is by the stepwise substitution of only one or two amino acid residues out of the total protein structure. Although a large database of sequence-structure correlations is available, and growing rapidly together with the necessary software, it is presently insufficient accurately to predict three-dimensional changes as a result of such substitutions. The main problem is assessing the long-range effects, including solvent interactions, on the new structure. As the many reported results would attest, the science is at a stage where it can explain the structural consequences of amino acid substitutions after they have been determined but cannot accurately predict them. Protein engineering, therefore, is presently rather a hit or miss process which may be used with only little realistic likelihood of immediate success. Apparently quite small sequence changes may give rise to large conformational alterations and even affect the rate-determining step in the enzymic catalysis. However it is reasonable to suppose that, given a sufficiently detailed database plus suitable software, the relative probability of success will increase over the coming years and the products of protein engineering will make a major impact on enzyme technology.Much protein engineering has been directed at subtilisin (from Bacillus amyloliquefaciens), the principal enzyme in the detergent enzyme preparation, Alcalase. This has been aimed at the improvement of its activity in detergents by stabilising it at even higher temperatures, pH and oxidant strength. Most of the attempted improvements have concerned alterations to:

1. the P_1 cleft, which holds the amino acid on the carbonyl side of the targeted peptide bond;
2. the oxyanion hole (principally Asn_{155}), which stabilises the tetrahedral intermediate;
3. the neighbourhood of the catalytic histidyl residue (His_{64}), which has a general base role; and
4. the methionine residue (Met_{222}) which causes subtilisin's lability to oxidation.

It has been found that the effect of a substitution in the P_1 cleft on the relative specific activity between substrates may be fairly accurately predicted even though predictions of the absolute effects of such changes are less successful. Many substitutions, particularly for the glycine residue at the bottom of the P_1 cleft (Gly_{166}), have been found to increase the specificity of the enzyme for particular peptide links whilst reducing it for others. These effects are achieved mainly by corresponding changes in the K_m rather than the V_{max}.

Increases in relative specificity may be useful for some applications. They should not be thought of as the usual result of engineering enzymes, however, as native subtilisin is unusual in being fairly non-specific in its actions, possessing a large hydrophobic binding site which may be made more specific relatively easily (e.g. by reducing its size).

The inactivation of subtilisin in bleaching solutions coincides with the conversion of Met_{222} to its sulfoxide, the consequential increase in volume occluding the oxyanion hole. Substitution of this methionine by serine or alanine produces mutants that are relatively stable, although possessing somewhat reduced activity. An example of the unpredictable nature of protein engineering is given by trypsin, which has an active site closely related to that of subtilisin.

Substitution of the negatively charged aspartic acid residue at the bottom of its P_1 cleft, which is used for binding the basic side-chains of lysine or arginine, by positively charged lysine gives the predictable result of abolishing the activity against its normal substrates but unpredictably also gives no activity against substrates where these basic residues are replaced by aspartic acid or glutamic acid. Considerable effort has been spent on engineering more thermophilic enzymes. It has been found that thermophilic enzymes are generally only 20-30 kJ more stable than their mesophilic counterparts. This may be achieved by the addition of just a few extra hydrogen bonds, an internal salt link or extra internal hydrophobic residues, giving a slightly more hydrophobic core. All of these changes are small enough to be achieved by protein engineering.

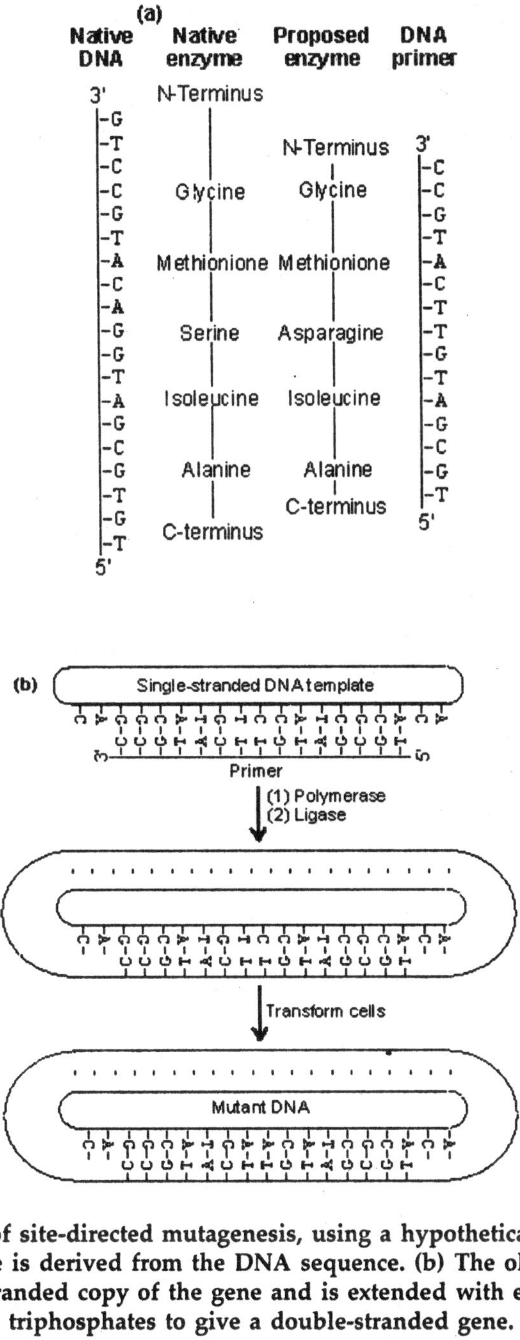

An outline of the process of site-directed mutagenesis, using a hypothetical example. (a) The primary structure of the enzyme is derived from the DNA sequence. (b) The oligonucleotide primer is annealed to a single-stranded copy of the gene and is extended with enzymes and nucleotide triphosphates to give a double-stranded gene.

To ensure a more predictable outcome, the secondary structure of the enzyme must be conserved and this generally restricts changes in the exterior surface of the enzyme. Suitable for exterior substitutions for increasing thermostability have been found to be aspartate → glutamate, lysine → glutamine, valine → threonine, serine → asparagine, isoleucine → hreonine, asparagine → aspartate and lysine → arginine. Such substitutions have a fair probability of success. Where allowable, small increases in the interior hydrophobicity for example by substituting interior glycine or serine residues by alanine may also increase the thermostability. It should be recognised that making an enzyme more thermostable reduces its overall flexibility and, hence, it is probable that the factitious enzyme produced will have reduced catalytic efficiency.

ARTIFICIAL ENZYMES

A number of possibilities now exist for the construction of artificial enzymes. These are generally synthetic polymers or oligomers with enzyme-like activities, often called synzymes. They must possess two structural entities, a substrate-binding site and a catalytically effective site. It has been found that producing the facility for substrate binding is relatively straightforward but catalytic sites are somewhat more difficult. Both sites may be designed separately but it appears that, if the synzyme has a binding site for the reaction transition state, this often achieves both functions.For a one-substrate reaction the reaction sequence is given by

$$\text{synzyme} + S \rightleftharpoons (\text{synzyme-S complex}) \rightarrow \text{synzyme} + P \quad [5]$$

Some synzymes are simply derivatised proteins, although covalently immobilised enzymes are not considered here. An example is the derivatisation of myoglobin, the oxygen carrier in muscle, by attaching $(Ru(NH_3)_5)^{3+}$ to three surface histidine residues. This converts it from an oxygen carrier to an oxidase, oxidising ascorbic acid whilst reducing molecular oxygen. The synzyme is almost as effective as natural ascorbate oxidases.It is impossible to design protein synzymes from scratch with any probability of success, as their conformations are not presently predictable from their primary structure. Such proteins will also show the drawbacks of natural enzymes, being sensitive to denaturation, oxidation and hydrolysis. For example, polylysine binds anionic dyes but only 10% as strongly as the natural binding protein, serum albumin, in spite of the many charges and apolar side-chains. Polyglutamic acid, however, shows synzymic properties. It acts as an esterase in much the same fashion as the acid proteases, showing a bell-shaped pH-activity relationship, with optimum activity at about pH 5.3, and Michaelis-Menten kinetics with a K_m of 2 mm and Vmax of 10-4 to 10-5 s-1 for the hydrolysis of 4-nitrophenyl acetate. Cyclodextrins (Schardinger dextrins) are naturally occurring toroidal molecules consisting of six, seven, eight, nine or ten α-1, 4-linked D-glucose units joined head-to-tail in a ring (α–, β–, γ–, δ- and ε-cyclodextrins, respectively:

they may be synthesised from starch by the cyclomaltodextrin glucanotransferase) from Bacillus macerans). They differ in the diameter of their cavities (about 0.5-1 nm) but all are about 0.7 nm deep. These form hydrophobic pockets due to the glycosidic oxygen atoms and inwards-facing C-H groups. All the C-6 hydroxyl groups project to one end and all the C-2 and C-3 hydroxyl groups to the other.

Their overall characteristic is hydrophilic, being water soluble, but the presence of their hydrophobic pocket enables them to bind hydrophobic molecules of the appropriate size. Synzymic cyclodextrins are usually derivatised in order to introduce catalytically relevant groups. Many such derivatives have been examined. For example, a C-6 hydroxyl group of β-cyclodextrin was covalently derivatised by an activated pyridoxal coenzyme. The resulting synzyme not only acted a transaminase but also showed stereoselectivity for the L-amino acids. It was not as active as natural transaminases, however. Polyethyleneimine is formed by polymerising ethyleneimine to give a highly branched hydrophilic three-dimensional matrix. About 25% of the resultant amines are primary, 50% secondary and 25% tertiary:

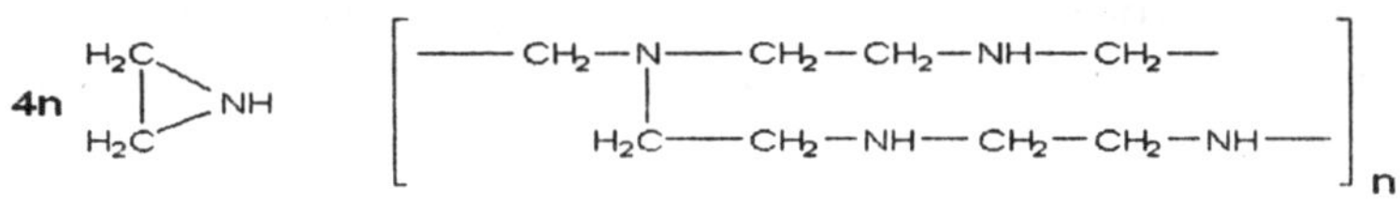

[6]

Ethyleneimine polyethyleneimine

The primary amines may be alkylated to form a number of derivatives. If 40% of them are alkylated with 1-iodododecane to give hydrophobic binding sites and the remainder alkylated with 4(5)-chloromethylimidazole to give general acid-base catalytic sites, the resultant synzyme has 27% of the activity of a-chymotrypsin against 4-nitrophenyl esters. As might be expected from its apparently random structure, it has very low esterase specificity. Other synzymes may be created in a similar manner.

Antibodies to transition state analogues of the required reaction may act as synzymes. For example, phosphonate esters of general formula $(R\text{-}PO_2\text{-}OR')^-$ are stable analogues of the transition state occurring in carboxylic ester hydrolysis. Monoclonal antibodies raised to immunising protein conjugates covalently attached to these phosphonate esters act as esterases.

The specificities of these catalytic antibodies (also called abzymes) depends on the structure of the side-chains (i.e. R and R' in $(R\text{-}PO_2\text{-}OR')^-$) of the antigens. The Km values may be quite low, often in the micromolar region, whereas the Vmax values are low (below 1 s^{-1}), although still 1000-fold higher than hydrolysis by background hydroxyl ions. A similar strategy may be used to produce synzymes by molecular 'imprinting' of polymers, using the presence of transition state analogues to shape polymerising resins or inactive non-enzymic protein during heat denaturation.

COENZYME-REGENERATING SYSTEMS

Many oxidoreductases and all ligases utilise coenzymes (e.g. NAD+, $NADP^+$, NADH, NADPH, ATP), which must be regenerated as each product molecule is formed. Although these represent many of the most useful biological catalysts, their application is presently severely limited by the high cost of the coenzymes and difficulties with their regeneration. These two problems may both be overcome at the same time if the coenzyme is immobilised, together with the enzyme, and regenerated in situ.

A simple way of immobilising/regenerating coenzymes would be to use whole-cell systems and these are, of course, in widespread use. However as outlined earlier, these are of generally lower efficiency and flexibility than immobilised-enzyme systems. Membrane reactors may be used to immobilise the coenzymes but the pore size must be smaller than the coenzyme diameter, which is extremely restrictive. Coenzymes usually must be derivatised for adequate immobilisation and regeneration. When successfully applied, this process activates the coenzymes for attachment to the immobilisation support but does not interfere with its biological function. The most widely applied synthetic routes involve the alkylation of the exocyclic N^6-amino nitrogen of the adenine moiety present in the coenzymes NAD^+, $NADP^+$, NADH, NADPH, ATP and coenzyme A.

In some applications, such as those using membrane reactors it is only necessary that the coenzyme has sufficient size to be retained within the system. High molecular weight water-soluble derivatives are most useful as they cause less diffusional resistance than insoluble coenzyme matrices. Dextrans, polyethyleneimine and polyethylene glycols are widely used. Relatively low levels of coenzyme attachment are generally sought in order to allow greater freedom of movement and avoid possible inhibitory effects. The kinetic properties of the derived coenzymes vary, depending upon the system, but generally the Michaelis constants are higher and the maximum velocities are lower than with the native coenzymes.

Coenzymes immobilised to insoluble supports presently have somewhat less favourable kinetics even when co-immobilised close to the active site of their utilising enzymes. This situation is expected to improve as more information on the protein conformation surrounding the enzymes' active sites becomes available and immobilisation methods become more sophisticated. However, the cost of such derivatives is always likely to remain high and they will only be economically viable for the production of very high value products. There are several systems available for the regeneration of the derivatised coenzymes by chemical, electrochemical or enzymic means. Enzymic regeneration is advantageous because of its high specificity but electrochemical procedures for regenerating the oxidoreductase dinucleotides are proving competitive. To be useful in regenerating coenzymes, enzymic processes must utilise cheap substrates and readily available enzymes and give non-interfering and easily separated products. Formate dehydrogenase and acetate kinase present useful examples of their use, although the presently available commercial enzyme preparations are of low activity:

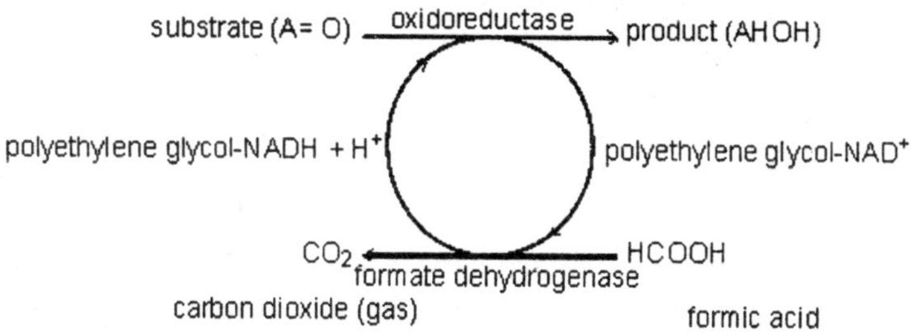

[7]

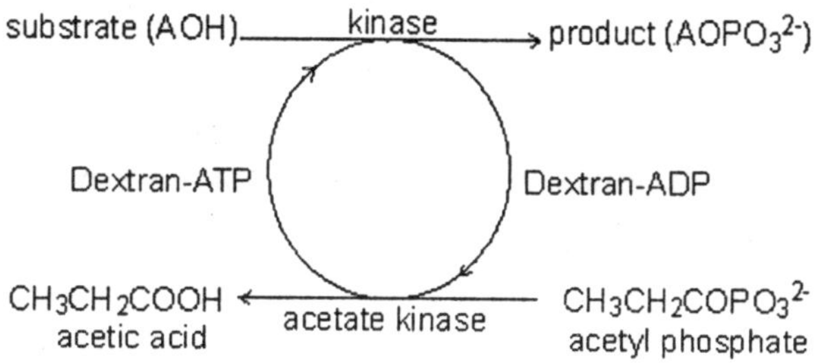

[8]

ADVANCES IN ENZYME TECHNOLOGY

Enzymic Reactions in Biphasic Liquid Systems

It would often be useful if enzyme catalysed reactions could be performed in solvents other than water, as this is not the ideal medium for the majority of organic reactions. Many reactants (e.g. molecular oxygen, steroids and lipids) are more soluble in organic solvents than in water and some products may be quite labile in an aqueous environment. Accomplishing reactions with such substrates (e.g. the aerobic oxidation of oestrogens catalysed by fungal laccase) in non-aqueous media allows a much increased volumetric activity to be achieved. Microbial contamination, by contrast, is much less of a problem in such solvents, and the consequent absence of microbial proteases may lead to an apparent stabilisation in the biocatalyst. Some polymerising reactions, for example the polymerisation of phenols catalysed by peroxidase, will produce a higher molecular weight product when carried out in a solution more able to dissolve the products (i.e. oligomers) initially formed. Under normal physiological conditions, hydrolytic enzymes catalyse the degradation of polymers; i.e. hydrolases are transferases normally transferring a moiety to the acceptor, water.

Water is normally present in a vast molar excess over other potential acceptor molecules so no reaction occurs other than hydrolysis. Also, the normal 'concentration' of water (about 55.5 M) is much greater than its typical Km (about 50 mM) and the rate of hydrolysis will not be affected as the reaction proceeds. By greatly reducing the water activity in these systems they can be used to transfer to other acceptors. Examples of this can be found in the transesterification reactions of esterases and lipases, described more fully later, and the (undesirable) formation of isomaltose from glucose catalysed by glucoamylase. Restriction of the enzyme to the aqueous phase effectively immobilises the enzyme and allows its straightforward separation, using phase separators developed for the established chemical process industry, from product-containing organic phase.

The main asset of these systems, however, is their ability to shift the thermodynamic equilibria of the reactions. Although changes in the physical conditions (i.e. temperature, pH and pressure) do affect the K_{eq} of a reaction, usually this effect is relatively slight over the physical range allowed by stability of the biocatalysts. Use of a biphasic aqueous-organic system, however, may result in substantial changes in the practically useful 'apparent' K_{eq}. The use of enzymes within organic solvents normally results in a two phase system as all water-soluble enzymes possess a significant amount of strongly bound water, even when in an apparently dry state.The surfactant (e.g. cetyltrimethylammonium bromide (CTAB), bis(2-ethylhexyl) sodium sulphosuccinate (AOT), phosphatidylcholine, tetraethyleneglycoldodecylether) is only found at the interphase boundary, whereas the water-immiscible cosurfactant (e.g. butanol, hexanol, octanol), added to vary the properties of this interphase, is generally less polar and more soluble in the organic continuous phase. Both preparations (a) and (b) give optically transparent solutions.

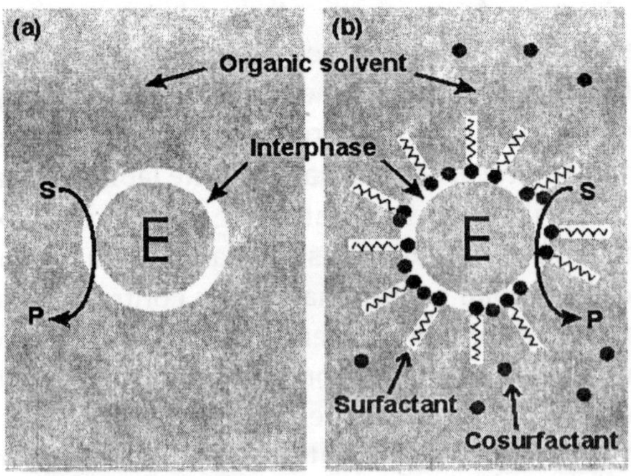

Schematic diagram showing two configurations for an enzyme within an organic solvent. (a) almost-anhydrous enzyme suspended in the organic solvent. The enzyme (E) is surrounded by a thin interphase consisting of water or water plus immobilisation support. (b) enzyme dissolved in a reversed micellar medium. The micelles are formed by the surfactant molecules with assistance from the cosurfactant (if present).

Stabilisation of Enzymes in Biphasic Aqueous-organic Systems

It should become clear from the later discussion that there may be a substantial advantage to be gained from the use of biphasic systems in many enzyme-catalysed reactions. One major factor must first be addressed; the stability of the enzyme in these systems. A distinction should be drawn between the more water-soluble hydrophilic enzymes and the more hydrophobic enzymes often associated with lipid and membranes (e.g. lipases). The active integrity and stability of hydrophilic enzymes appears to depend on the presence of a thin layer of water, just a few molecules thick, within the microenvironment. This amount of water is miniscule (between 50 and 500 molecules of water for each enzyme molecule) and the enzyme may effectively be operating in an almost anhydrous state. Some hydrophobic lipases retain activity even if fewer molecules of water remain; presumably just sufficient to stabilise the conformation of the active site.

The pH of such minute pools of water, containing no free hydrogen ions, is impossible to measure, or control, directly. However, it appears that the enzyme 'remembers' the pH of its last aqueous solution and functions as though at that pH. If the enzyme-bound water is stripped out or diluted by the use of the more water-soluble, or miscible, organic solvents then the enzyme is usually inactivated. However, under conditions where this does not occur, the limited amount of water available, and the associated reduction in the water activity, considerably reduces the rate of thermoinactivation. This has a stabilising effect on most enzymes; porcine pancreatic lipase, for example, has a half-life of greater than 12 hours at 100°C in 0.02% water in tributyrin, whereas this drops to 12 minutes at a 0.8% water content and inactivation is almost instantaneous in 100% water. Additionally, the freezing point of the water is reduced which allows the use of particularly heat-labile enzymes at very low temperatures. The lowering of the water activity tends to produce a more rigid enzyme molecule which may affect both its K_m and V_{max}. In extreme cases, this may result in a change in the catalytic properties. Porcine pancreatic lipase demonstrates this effect. When used in biphasic systems of low water activity, it no longer catalyses transesterification reactions involving the bulky tertiary alcohols.

The most important factor in the balance between stabilisation and inactivation, due to organic phase, is the solvent polarity. Solvents of lower polarity (i.e. greater hydrophobicity) are less able to disrupt the structure of the necessary tightly bound water molecules. The best measure of polarity is the logarithm of the partition coefficient (Log*P*) of the organic liquid between n-octanol and water; the higher the Log*P*, the more non-polar (hydrophobic) is the solvent (Table 1).

$$\text{Log}\boldsymbol{P} = \text{Log}_{10}\left(\frac{[Material_{octanol}]}{[Material_{water}]}\right) \qquad [9]$$

Table 1: LogP values of the more commonly used organic solvents.

Solvent	LogP	Solvent	Log*P*
Butanone	0.3	1,1,1-trichloroethane	2.8
Ethyl acetate	0.7	Carbon tetrachloride	2.8
Butanol	0.8	Dibutyl ether	2.9
Diethyl ether	0.8	Cyclohexane	3.1
Methylene chloride	1.4	Hexane	3.5
Butyl acetate	1.7	Petroleum ether (60-80)	3.5
Di-isopropyl ether	2.0	Petroleum ether (80-100)	3.8
Benzene	2.0	Dipentyl ether	3.9
Chloroform	2.2	Heptane	4.0
Tetrachloroethylene	2.3	Petroleum ether (100-120)	4.3
Toluene	2.7	Hexadecane	8.7

Log*P* values increase by about 0.52 for every methylene group ($-CH_2-$) added in an homologous series. Thus, the Log*P* of hexanol is that of butanol (0.8) plus 2 x 0.52 (i.e. approximately 1.8). There appears to be a clear correlation between the activity of biocatalysts in two-phase systems and the LogP. The S-shape of this relationship suggests that enzymes are generally inactivated by solvents with log*P* < 2 but are little affected by solvents with Log*P* > 4. There is some variation in the effects between different enzymes and different solvents which makes activity prediction, in the Log*P* range of 2 - 4, rather imprecise.

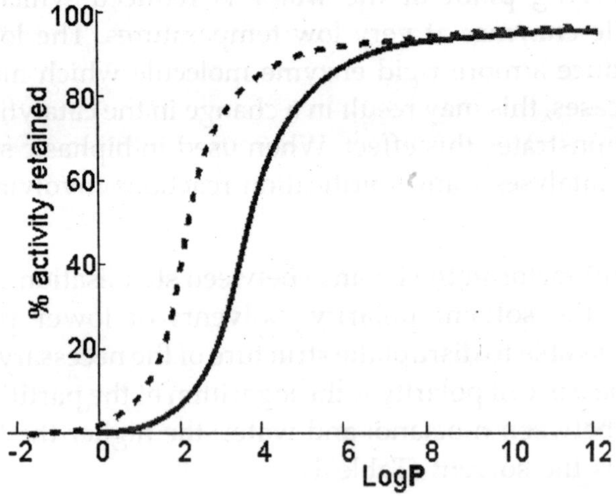

Schematic diagram showing the dependence of the activity of immobilised enzymes, in biphasic systems, on the LogP of the organic phase. — free enzyme; -------- enzyme immobilised within a strongly hydrophilic support.

This range includes some of the most utilised organic phases (e.g. chloroform), which may be suitable for some applications but cause harmful inactivation in others. The solubility of the reactant(s) and product(s) may considerably reduce the range of LogP that are available for a particular application; many basically nonpolar molecules possessing some polar structural regions which cause their lack of solubility in strongly hydrophobic solvents.

The choice of organic phase will also depend on additional factors such as cost, ease of recovery, fire and fume hazards and specific inhibitory effects. The S-shaped curve can be shifted to the left by immobilising the enzyme within a highly hydrophilic support. A simple way of achieving this is to impregnate a beaded hydrophilic polymer (e.g. Sephadex, agarose) with the enzyme followed by suspension of the wet beads directly in the organic phase. These shifts are very important as they greatly increase the choice of suitable organic phase. Such impregnated beads have the additional advantages that

a. they protect the contained enzyme from liquid-liquid interfacial denaturation at higher rates of stirring;

b. they enable more facile recovery of the biocatalyst;

c. they may be used with low molecular weight hydrophilic coenzymes (e.g. $NAD(P)^+$) with the assistance of a coenzyme regenerating process; the coenzyme being effectively immobilised within the aqueous pools; and

d. they allow the efficient and continuous use of biphasic PBRs, so long as the moving organic phase remains saturated with water.

Biphasic systems may be further stabilised by the use of deuterated water (D_2O). This reduces the rate of thermal inactivation, although it does cause an increase in the pK_as of ionising groups by about 0.4 with the associated changes in the pH-activity and pH-stability relationships. The higher cost of the deuterated water is offset, to a certain extent, by the small amount necessary in these systems and the ease with which it may be recovered at the end of a catalytic process. Even where the organic solvent has very low Log*P* and is miscible, the effect of the expected loss in enzymic activity may be offset by changes in the equilibrium constant. Thus it has been proposed that glucose isomerase be used in aqueous ethanol to produce high fructose corn syrup. The equilibrium fraction of fructose can be raised by about 10% to 55% at 30°C in 80% v/v ethanol. This is economically valid even though the enzymic activity drops by about 10% compared to that in the absence of ethanol.

Enzyme Kinetics in Biphasic Aqueous-organic Systems

Although the equilibrium position of reactions may be shifted in biphasic systems, this is of no practical consequence unless the reaction actually occurs. The enzyme must be able to convert the reactants to products. For this to proceed, even in the presence of fully active enzymes, there must be some reactant present in the aqueous microenvironment surrounding the enzyme. Reactants may have fairly low aqueous solubilities under the normal conditions associated with such biphasic processes. The aqueous phase remains effectively unstirred, even when the organic phase is well-stirred, during the reaction so that the thickness of the 'unstirred layer' (d) is equal to the thickness of the aqueous layer surrounding the enzyme. Where d is large, the rate of reaction is likely to be controlled by the passage of reactants through this aqueous layer as the concentration gradients will be very shallow due to the low solubilities of the reactants in water. However, d is often very thin, however, especially where the enzyme is freely soluble, causing the reactions to be effectively diffusionally controlled by the passage of the reactants from the more concentrated organic phase through the interphase boundary to very low concentrations within the microenvironment of the enzyme.

Clearly, the thickness of the aqueous layer around an immobilised enzyme will be of great importance in determining the degree of diffusional resistance involved, and hence the lowering of the reaction rate. As with all immobilised enzymes, the volumetric surface area (A/V) is an important parameter governing the overall flux of the substrate to the biocatalytic surface, the effectiveness of the enzyme and the productivity. Whereas in reactions involving a single liquid phase this surface area is the area presented by the exterior of the particles to the bulk of the liquid, in most practical biphasic liquid systems it is the interfacial area between the two liquid phases that is relevant.

Under normal operating conditions this interfacial area will depend upon the volume of the aqueous droplet surrounding and enclosing the enzymic biocatalyst. As the relative volume of the organic phase increases the volumetric surface area will decrease, so reducing the maximum volumetric productivity. This will be of particular importance where a needs to be very high in order to shift the reaction equilibrium. Clearly this will reduce the reaction rate due to the reduction in the amount of enzyme that can be present, due to its associated water, and the associated reduction in the interfacial area. Where the catalysed reaction generates water, this should not generally be allowed to accumulate as it not only has a detrimental effect on the equilibrium constant but will also slow down the rate of reaction.

There are some rules which allow the optimisation of the two-phase system for enzyme-catalysed reactions. They depend upon the concept of Log*P* being extended to cover the substrates, products and the interphase.

1. The difference between the Log*P* values of the substrate(s) and the interphase should be as small as possible whereas that of the substrate(s) should be much lower than that of the organic phase. These conditions encourage high concentrations of substrate(s) within the interphase and, hence, the transfer of the substrate from the organic phase into the aqueous phase.

2. The difference between the Log*P* values of the product(s) and the organic phase should be as small as possible whereas that of the product(s) should be much greater than that of the interphase. These conditions encourage the transfer of the product from the aqueous phase through the interphase into the organic phase, after reaction.

The Log*P* of the interphase can be varied independently of the organic phase by suitable choice of surfactants and cosurfactants. When there is excess water and surfactant present, the surfactant may form multiple concentric membranes, consisting of surfactant bilayers, around the aqueous core. This presents another reason for the removal of the water produced by use of hydrolases (used as synthetases) in biphasic systems. Control over the amount of water surrounding the enzyme is often possible by means of molecular sieves (e.g. potassium aluminium silicate), which absorb water. Removal of water dissolved in the organic phase can easily be achieved by their use and this leads to the depletion of water from the aqueous biocatalytic pools by its partition. If the organic phase has a high boiling point the removal of water may be simply achieved by vacuum distillation; e.g. in the esterification of fatty acids and fatty alcohols catalysed by lipase and used in wax manufacture.

Use of Aqueous Two-phase Systems

Aqueous 2-phase systems offer the opportunity to shift reaction equilibria towards product formation by ensuring that enzyme and substrate partition into one phase whilst the product enters, and may be removed from, the other. The theory developed (earlier) for aqueous/organic biphasic systems is equally applicable to these systems. An example of such an extractive bioconversion is a method for the conversion of starch to glucose using bacterial a-amylase and glucoamylase. Starch substrate partitions almost entirely into the lower, more hydrophilic, dextran-rich phase of a system comprising 3% polyethylene glycol and 5% crude unfractionated dextran. The enzymes also partition largely to the bottom phase but the glucose, produced by the hydrolysis, distributes itself more evenly between the phases. A small proportion of the glucoamylase enters the upper phase and will convert any oligosaccharide entering that phase to glucose. Concentrations of up to 140 g l^{-1} glucose may be reached in the upper phase.Such a system offers some of the advantages of an immobilised enzyme process without some of the disadvantages. The enzymes are largely retained and are stabilised by the presence of the polymers yet catalysis is in homogeneous solution (within the phase) so no diffusion limitations to mass transfer exist.

Practical Examples of the Use of Enzymes 'in Reverse'

It has long been known that if proteases are supplied with high concentrations of soluble proteins, peptides or amino acids, polymers (plasteins) are produced with apparently random, if rather hydrophobic, structures. This reaction has been used, for example, to produce bland-tasting, colourless plasteins from brightly coloured, unpleasant tasting, algal biomass and for the introduction of extra methionine into low quality soy protein. However in general, the non-specific use of proteases in the synthesis of new structures has not found commercial use. However, proteases have come into use as alternatives to chemical methods for the synthesis of peptides of known and predetermined structure because their specificity allows reactions to proceed stereospecifically and without costly protection of side-chains.

A method for the synthesis of the high intensity sweetener aspartame exemplifies the power of proteases (in this case, thermolysin). Aspartame is the dipeptide of L-aspartic acid with the methyl ester of L-phenylalanine (α-L-aspartyl-L- phenylalanyl-O-methyl ester). The chemical synthesis of aspartame requires protection of both the β-carboxyl group and the α-amino group of the L-aspartic acid. Even then, it produces aspartame in low yield and at high cost. If the β-carboxyl group is not protected, a cost saving is achieved but about 30% of the β-isomer is formed and has, subsequently, to be removed. When thermolysin is used to catalyse aspartame production the regiospecificity of the enzyme eliminates the need to protect this β-carboxyl group but the a-amino group must still be protected (usually by means of reaction with benzyl chloroformate to form the benzyloxycarbonyl derivative, i.e. BOC-L-aspartic acid) to prevent the synthesis of poly-L-aspartic acid. More economical racemic amino acids can also be used as only the desired isomer of aspartame will be formed.

If stoichiometric quantities of BOC-L-aspartic acid and L-phenylalanine methyl ester are reacted in the presence of thermolysin, an equilibrium reaction mixture is produced giving relatively small yields of BOC-aspartame. However, if two equivalents of the phenylalanine methyl ester are used, an insoluble addition complex forms in high yield at concentrations above 1 M. The loss of product from the liquid phase due to this precipitation greatly increases the overall yield of this process. Later, the BOC-aspartame may be released from this adduct by simply altering the pH. The stereospecificity of the thermolysin determines that only the L-isomer of phenylalanine methyl ester reacts but the addition product is formed equally well from both the D- and L-isomers.

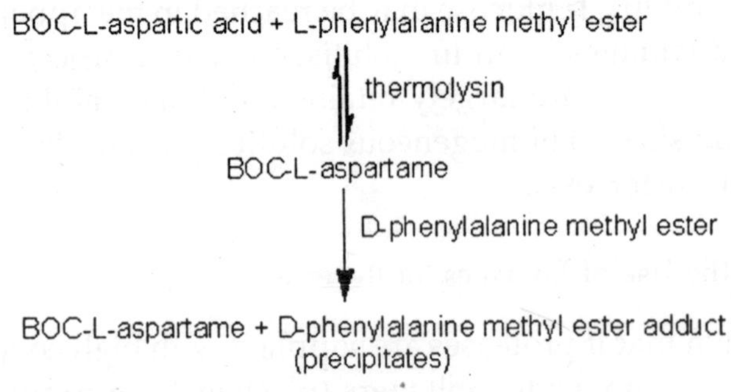

[9]

This fortuitous state of affairs allows the use of racemic phenylalanine methyl ester, the L-isomer being converted to the aspartame derivative and the D-isomer forming the insoluble complex shifting the equilibrium to product formation. D-phenylalanine ethyl ester released from the addition complex may be isomerised enzymically to reform the racemic mixture. The BOC-aspartame may be deprotected by a simple hydrogenation process to form aspartame. Immobilised thermolysin cannot be used in this process as it has been found to co-precipitate with the insoluble adduct. However, this may be circumvented by its use within a liquid/liquid biphasic system.

Glycosidases used in Synthetic Reactions

This is a comparatively neglected topic probably because polysaccharides can be obtained readily from plant or microbial sources, because polysaccharide function is not so specifically related to its structure, and because the theory describing the functional significance of oligosaccharides is still being developed. As there is an abundance of knowledge about glycosidases and their specificities, there is no fundamental reason why they should not be used to synthesise oligosaccharides.

There are few examples in the literature of oligosaccharides being constructed using enzymes such as dextransucrase, levansucrase, cyclomaltodextrin glucanotransferase and b-galactosidase to build up glucose, fructose or galactose residues on suitable acceptors. Fructosyltransferases have been used to synthesise compounds such as sucrose-6-acetate, xylosucrose (β-D-fructofuranosyl-(2,1)-α-D-xylopyranoside) and the low cariogenic sweetener, 'neosugar' which consists of a mixture of glucose, sucrose and β-2,1 linked fructans with terminal non-reducing glucose residues (typical composition before chromatographic refinement: 2% fructose, 26% glucose, 11% sucrose, 30% 1-kestose (β-D-fructofuranosyl-(2,1)-β-D-fructofuranosyl- (2,1)-α-D-glucopyranose), 25% nystose (β-D-fructofuranosyl- (2,1)-β-D-fructofuranosyl-(2,1)-β-D-fructofuranosyl-(2,1)-α-D- glucopyranose) and 6% higher oligosaccharides).

$$\text{sucrose} + \text{glucose-6-acetate} \rightleftharpoons \text{sucrose-6-acetate} + \text{glucose} \qquad [10]$$

$$\text{sucrose} + \text{xylose} \rightleftharpoons \text{xylosucrose} + \text{glucose} \qquad [11]$$

$$\text{sucrose} + \text{sucrose} \rightleftharpoons \text{1-kestose} + \text{glucose} \qquad [12]$$

$$\text{1-kestose} + \text{sucrose} \rightleftharpoons \text{nystose} + \text{glucose} \qquad [13]$$

As shown in reaction scheme [14], dextransucrase can be used to produce dextran (a polymer of a-1,6 linked glucose residues) from sucrose. Dextran is used in gel chromatographic media, such as Sephadex, in the aqueous 2-phase systems already described, and as a blood plasma extender. For any of these uses it is undesirable to have cells remaining in the dextran so the dextransucrase, produced extracellularly by Leuconostoc mesenteroides is purified before use. The size of dextran molecules produced may be controlled by including small concentrations of sugars such as maltose which compete with sucrose as acceptors for glucose residues from the donor sucrose molecules.

$$\text{sucrose} + \text{dextran } (G_n) \rightleftharpoons \text{fructose} + \text{dextran } (G_{n+1}) \qquad [14]$$

Thermodynamically controlled synthesis of oligosaccharides, catalysed by glycosidases, is possible by use of high substrate concentrations and a 'molecular trap' to remove the products. For example, β-galactosidase may be used to synthesise N-acetyllactosamine (β-D-galactopyranosyl-(1,6)-2-acetamido- 2-deoxy-D-glucose) using a carbon-celite column to remove it, as formed.

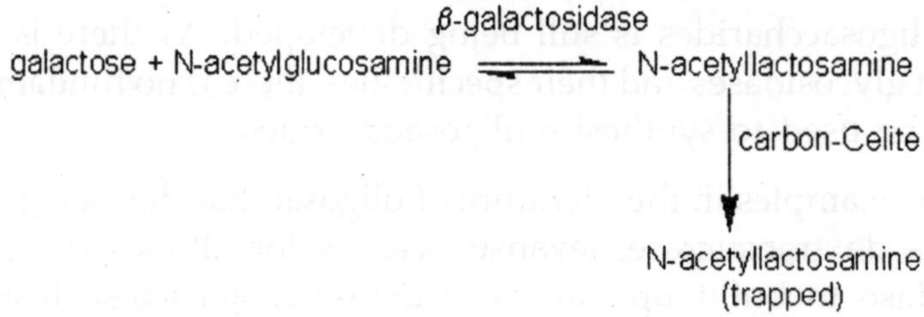

[15]

There seems to have been no systematic attempt to construct hetero-oligosaccharides in a stepwise movement yet thermodynamically or kinetically controlled reactions could be used just as in peptide synthesis.

REFERENCES

Acerenza, L. & Kacser, H., "Enzyme kinetics and metabolic control. A method to test and quantify the effect of enzymic properties on metabolic variables", *Biochem. J.*, 1990.

Flint, H. J., Dible, S. & Kacser, H., "Derepression of enzyme synthesis is response to arginine limitation in *Neurospora crassa*", *J. Gen. Microbiol.*, 1985.

Kacser, H., "The control of enzyme systems *in vivo*", *Elasticity analysis of the steady state, Biochem. Soc. Trans.*, 1983.

Middleton, R. J. & Kacser, H., "Enzyme variation, metabolic flux and fitness: alcohol dehydrogenase in *D. melanogaster*", *Genetics*, 1983.

Sauro, H. M. & Kacser, H., "Enzyme-enzyme interactions and control analysis", 2. The case of non-independence: heterologous associations, *Eur. J. Biochem.*, 1990.

7

Molecular Biology of Plant Genetic Engineering

Only some 10% of the world's 13 billion ha is farmed, although one third of the total land area is considered as potentially suitable for arable agriculture to some degree. Even so, abiotic stress in one form or another, still limits production on most of the world's 1.4 billion farmed hectares. This is a problem that is not going to go away. For example, yield reductions due to drought stress are already serious, and they will increase. Irrigation will cease to be a practical solution as water becomes scarcer, and the irrigation already in place will continue to lead to yet more soil salinisation. High and low temperatures, acid soils and soils with high levels of metal ions continue to reduce productivity over vast tracts of land and will remain an agricultural challenge for the foreseeable future.

Solutions to the problem will be as diverse as the lands affected. However new, locally adapted and improved varieties will always be a central component in any package of engineering, agricultural management, sociological and political solutions. Moreover in these times of surpluses in developed countries, solutions to the problem of abiotic stress are laid firmly at the door of developing country agriculture. It is here that the most severe stresses are found and here the need for increased food production to feed an increasing population is greatest.

The Consultative Group on International Agricultural Research (CGIAR) was originally created at the initiative of the Rockefeller Foundation, which had sponsored international meetings of agronomists at its Bellagio Conference Center in Lake Como, Italy, from 1968 onwards. In 1970, foundation officials proposed a worldwide network of agricultural research centers under a permanent secretariat. This was further supported and developed by the World Bank; on May 19, 1971, with the FAO, IFAD and UNDP as co-sponsors, the CGIAR was established. By 1983 there were thirteen research centers around the world under its umbrella.CGIAR now has 64 governmental and nongovernmental members and 15 research centres.At the time of its establishment there was widespread concern that developing countries would succumb to famine; the successes of the Green Revolution had started in Asia and the Pearson Commission on International Development had urged that the international community undertake "intensive international effort" to support "research specializing in food supplies and tropical agriculture". CGIAR was formed for the coordination of international agricultural research with the goals of poverty reduction and achieving food security in developing countries through agricultural research.The significance of abiotic stress has not been lost on CGIAR plant breeders. There is considerable work aimed at stress tolerant crop improvement already going ahead. Together with NARS and ARIs, the Centers are working towards an understanding of the genetic and physiological control of tolerance to the key stresses in their regions for their mandate crops, and are beginning to apply the results in breeding programmes. Progress, albeit incremental, is real and demonstrates that the problem is tractable to a genetic approach. In short, breeding is a viable option.

However developments elsewhere tell us that the future will not be the same as the past. Our science is becoming more generic on one hand, and more expensive on the other. The pressure is building for more centralisation. The science of abiotic stress resistance in the CGIAR could be the test-bed for a new way of working. Genetics itself is one such area, where a new science is emerging from the masses of DNA sequence and the associated novel and high throughput technologies. This new 'genomics' promises more rapid and more spectacular returns, but with expensive equipment, much of which has a short 'shelf life'. Some of these massively parallel genomics and gene manipulation technologies are already, and with some success, being turned on the abiotic stress tolerance problem in 'model' organisms by researchers at ARIs in developed countries. Some Centers are already tooling-up for plant genomics research. Another development is the discovery of 'synteny', where genome organisation has been found to be much more conserved over species than was previously thought. Application of synteny will allow advances in our knowledge about stress tolerance and the underlying genetics to be transferred between crop species. Synteny will similarly allow CGIAR and NARS scientists to apply the array of genomics resources already available in the models arabidopsis and rice to their mandate crops.

However, in order to harness synteny and the new genomics in a cost-effective and efficient manner it will be necessary to develop new ways of doing science. These could involve: more rationalisation and centralisation and sharing of expensive and rapidly improving technologies; more outsourcing to providers of standard scientific services and the sharing of skills by assembling multidisciplinary teams and networks in virtual centres that will work on a range of crop species. DNA science and the expensive equipment it begs, is identical for all organisms. Suddenly there is obvious potential for economies of scale in major collaborations.

The time could be right for a full-blooded assault on abiotic stress. Ongoing work shows that the motivation is already there. The question is not whether the work is needed, rather only when, how and what firepower should be brought to bear. A really effective collaboration will involve: the NARS with their germplasm collections, their knowledge of and, access to, stressed agricultural environments, and their plant breeders; ARIs with their experience of technology and model systems; the CGIAR Centers with their comparative advantage with the mandate crops, their collections and their networks to the developing world; and possibly industry as well.Many of these ideas have already been incorporated in a Global Challenge project, 'Unlocking genetic diversity in crops for the resource poor', and, apart from a recommendation to compile lists of potential alternative crops for use in sub-optimal soils and climates, are not dealt with further at length.Optimal organisation of genomics science within the System is relevant and not dealt with elsewhere.

The conclusions are that an increasing amount of work will be outsourced to specialist companies, leading ARIs or other Centers. There will likely be strong financial and infrastructural reasons for centralising other technologies, possibly micro-arrays today and soon the next generation of high throughput genotyping for marker-aided selection and germplasm characterisation. There are also scientific reasons for sharing intellectual resources that are in short supply or unevenly spread around the System, such as bioinformaticists and physiologists. There is an opportunity to cost-effectively appoint a central Genomics Facilitator who will carry out market testing, organise key facilities and link groups of researchers around the System so that we can best exploit the new generic aspects of our science.

MOLECULAR BIOLOGY AND GENETIC ENHANCEMENT OF CROP TOLERANCE

Only some 10% of the world's 13 billion ha is farmed. Apart from urban areas much of the remaining 11.5 billion ha are lands too hostile for any sort of agriculture. Moreover almost all the land that is farmable is under conditions sub-optimal, often to a considerable degree, for plant growth. Alongside losses due to pests and diseases, a further 70% of yield potential has been calculated to be lost to unfavourable physiochemical environments, even in developed agricultures. It is acknowledged that, in order to feed the eight billion mouths we expect by 2030, we will need to double world food production yet again. One component of that achievement will be the breeding of new varieties of food crops that will both improve yields on land presently being farmed on sub-optimal soils and extend our productive agriculture into lands which are currently barren.

Abiotic Stress

Drought

Unpredictable drought is the single most important factor affecting world food security and the catalyst of the great famines of the past. Moreover, because the world's water supply is fixed, increasing population pressures will ensure that the effects of successive droughts are more severe because competition from industry will increasingly limit the water available for agriculture. Crops are voracious consumers, for example, for paddy rice 5000 l of water is needed to produce 1 kg of grain. At present an unsustainable 70% of the world's water is used for agriculture. By 2025 it is expected that most Asian countries will join those that already have water shortages. Uncertainties over global warming raise yet further concerns. Drought stress is a concern for most crops at most Centres for most regions. These include, IITA Cowpea in the Sahel, soybean and tropical maize in the Dry Savanna, ICRISAT Sorghum, pearl millet, chickpea, groundnut and pigeon pea, CIAT Bean in Mexico, C America and NE Brazil, IRRI Rice in Bangladesh, E India, Thailand and Indonesia. CIP, Potatoes in China, India, Southern Africa, Kazakhstan and Afghanistan. CIMMYT, Wheat in C and W Asia and N Africa and maize in sub-Saharan Africa. ICARDA All crops (except faba bean which is only grown under irrigation) in N Africa and Asia. ICRISAT all crops in India and the Sahel.

Salt

Some 380 million ha, almost a third of the area farmed, is affected by salt, and the associated water logging and alkalinity. Sixty million ha are a direct result of over-irrigation, where a raised water table brings underground salt, particularly NaCl, to the surface. It is probable that this agricultural salinisation now degrades as much land as is put under new irrigation each year. Pressures on water use will ensure that the net productive irrigated land will go negative very soon and that secondary salinisation will become critical in Asia, Africa and S America.Salt stress is of particular significance for rice. IRRI Coastal salinity in Bangladesh, Orissa, Vietnam, Philippines and inland salinity in the Indogangetic plain and Thailand. ICARDA Secondary salinisation is a problem for all crops in C Asia.

Acid and degraded soils

Some 40% of the world's arable land is associated with acid soils, with pH less than 5, where growth is hindered by high aluminium or manganese content. This is particularly important in S America where some 380 million ha are affected, including almost the whole of the Amazon basin. Other excess metal ion contents reduce the agricultural potential of other soils. For example iron toxicity is a major problem affecting rice production in W Africa.Acid soils are a widespread problem. IITA Cowpea and soybean in the humid rain forest. CIAT Bean in Africa and both bean and *Brachairia* in L America. IRRI Rice in Bangladesh, Indonesia and Philippines. CIMMYT maize in L America, SE Asia and Africa. Wheat in CWANA.Other metal toxicities and deficiencies. IITA Low P for soybean. CIAT Low P for bean and *Brachairia*. IRRI Zn, P deficiency in Bangladesh, Indonesia and the Philippines and Fe deficiency in Sri Lanka and the Philippines. WARDA Fe deficiency is widespread in Africa. CIP Low P in China, Africa and in the Andes.

Low and high temperatures

Temperature also limits the range and production potential of many of our crops, even at tropical latitudes. Occasional and unpredictable periods of low temperature can be devastating to yields. For example, in the Andes 70% of land devoted to potato production is prone to cold stress.Cold stress is a rice problem for IRRI in Korea and Nepal. CIP, potatoes in the Andes. ICARDA Low temperature tolerance has become a problem associated with the shift from spring to autumn sowing for barley, lentils and chickpeas. Excessive heat is a problem for cowpea. IITA in the Sahel. CIP, for potatoes in S Asia.In fact, abiotic stress tolerance, particularly drought, is the priority target trait for most of the CG Centers dealing with crop plants. In the present economic and agricultural climate, with food surpluses in developed countries, the focus of the private sector will continue to be protection from disease and improvements in aspects of quality. Even given the extent of the problem and although abiotic stress is a significant factor for production in developed countries, it is unlikely that genetic solutions will be actively sought by commercial breeding companies. If the problem is to be tackled at all, abiotic stress tolerance mechanisms and their genetic application in the crops of the developing world will have to be addressed by the public sector working in the developing world.

Gene Mapping and Marker Development

Genetic mapping as a prerequisite to genetic analysis is now part of standard plant breeding. Those that are the focus of international effort, e.g. rice, wheat, potatoes, can use the well-developed public maps. Base maps for many of the 'orphan crops', in which there is little international trade, have been made at Centers or in Center-ARI collaborations. Only a few very minor mandated species remain unmapped.

The mapping of quantitative traits where there is often little knowledge of the genetic control in advance of the analysis, such as is usually the case with stress tolerance, is usually carried out by 'QTL mapping'. This requires a scan of the genome, with markers every 10 cM or so to identify those regions where segregation of the trait is associated with segregation for the markers. The reason much denser base maps are needed is that only a subset of the available markers will segregate in any single population. These locations are the basis for establishing a marker aided selection (MAS) breeding programme for tolerance and for eventual map-based cloning of the genes underlying the QTLs. Breeders' markers that are closely linked to the target gene may be derived straight from the base molecular map. Today the ideal marker system will be micro satellites, also known as simple sequence repeats (SSRs), although over the next few years single nucleotide polymorphisms (SNPs), which are more amenable to high throughput methods, will take over as the ideal marker.For rice, which will soon have the benefit of a full genome sequence, markers will never be a problem again. The sequence has been found to contain some 40,000 SSRs and SNPs and, base pair deletions or insertions indels, are found in unique sequence at a rate of about 1%, which works out at about 24 in every gene! However, apart from in the major cereals, an adequate supply of good quality markers for all applications is still a problem for most CGIAR mandate crops.

Genetic and Physiological Mechanisms

The physiological mechanisms underlying crop responses to stress and potential biochemical, physiological and architectural modifications that will allow crops to escape, avoid or tolerate stress are the subject of a vast literature. Two general approaches are taken in relation to varietal improvement, and both have their place. The 'empirical' approach proceeds from genotypic differences associated the best sources of tolerance in the cultivated crop or its wild relatives. Typically sources of tolerance are identified and then the underlying genetic control is investigated by QTL analysis in lines segregating for high and low tolerance. Although the identification of a predominant causal physiological mechanism is helpful, transfer of the improved trait to an already otherwise adapted variety can proceed simply for selecting for and accumulating 'beneficial' alleles. The second approach is often described as 'ideotype' breeding, in which specific morphologies or physiologies that might be expected to contribute to improved performance under stress are identified in diverse cultivated or wild germplasm and transferred to otherwise adapted varieties. The crossbreeding and, these days, marker-aided pyramiding of the underlying alleles is progressed in the same way in both approaches.

There is already considerable work underway at all CGIAR Centers to improve their mandate crops for stress tolerance. Almost all these breeding projects are being carried out in collaboration with NARS to address the major problems affecting their own agricultures. However many of these projects are crop- or geographical area-specific, even though the target tolerance and the technologies used to address the problem beg a collaborative, pan-stress, pan-crop, pan-Center approach.For example drought, salt stress and cold temperature stress are all physiologically linked because all three stress environments result in limiting the crops' physiological access to water. Thus many of the strategies for improved tolerance are likely to be multiply applicable. These will include osmotic adjustment in roots and leaves to retain water, erecting hydrophobic barriers in roots and leaves to retain water, and improving aquaporin efficiency to speed water movement in the plant. Although tolerance mechanisms might be expected to overlap, escape or avoidance mechanisms are more likely to be stress specific. For example reducing time to flowering may escape late season drought but will not help in a chronic saline situation. Deeper roots may be able to reach the last of the water in a drought but would only aggravate salt stress where the salt is being brought to the surface by a rising water table.

A few stress tolerances do usually appear to be under the control of major genes, as revealed by genetic analysis. Submergence tolerance is a prime example, and it may be no coincidence that this trait lends itself to a straightforward and definitive selection screen. This is a key trait in SE Asia where some 25 million ha are prone to flash flooding which can completely submerge the rice crop for several days. Here a single locus, *Sub1*, has been shown to provide substantial tolerance.Mention must be made of the potential of wild relatives as potential donors. Wild species, where the *raison d'être* is survival rather than yield, are likely to retain useful variation that may have been bred out of the cultivated crop. There are many examples where genes for tolerance have been identified in wild relatives and have been used to transfer useful variation to cultivated crops. In some cases the wild species themselves have been used directly to create new crop species, e.g. Tritipyrum incorporating salt tolerant *Thinopyrum bessabaricum*. The germplasm collections, which are mostly still relatively uncharacterised, will be central to any future stress tolerance initiative.

A possible alternative to varietal improvement is crop replacement. This is probably a viable socio-economic strategy only in extreme stress environments. Nevertheless NARS and their extension services should have reliable information about those crops which generally perform best in stress environments. These differences are exemplified by investigations into the responses of tropical grasses to aluminium stress. Signalgrass (*Brachairia decumbens*) was found to be far more tolerant than both close relatives and Al-resistant varieties of wheat, triticale and maize. It is probable that world-wide multi-centre, multi-crop studies, which are unlikely to be carried out by 'one-crop specialists', might reveal interesting alternatives in many stress situations, possibly even pushing back the borders of lands presently considered to marginal for agriculture at all.

Overall the genetic dissection of stress tolerance for developing countries is receiving considerable attention, particularly in crops that already have advanced genetic maps. Initial understanding of the physiological and genetic controls most certainly informs breeding programmes. Marker-aided pyramiding of several genes is probably the only way forward for the transfer of improved phenotypes that have been shown clearly to be under control of multiple loci. However these methods have yet to impact stress tolerance breeding programmes. Although a few endogenous genes have been identified which are likely to have major beneficial effects when used as a transgene, these have not yet been applied to practical breeding.

Mapping by NARS and CGIAR Centres has identified a number of genes, usually as anonymous QTLs. Most of these genes have not yet been associated with physiological mechanisms. Linkages between geneticist-breeders and physiologists could now pay dividends. The potential is there for application of these QTLs through MAS but this has not yet been generally successful. There is considerable scope for more collected wild and cultivated germplasm characterisation for stress tolerance. There is also a need to identify any potential alternative crops for severe or chronic stress environments. Any

Systemwide initiative to quantify crop yield potential under stress should use common genotypes across stresses and regions.

PROGRESS IN BREEDING

Some progress has been made in breeding for drought, salt and aluminium tolerance or avoidance. The CGIAR Center breeding programmes have played a major role in these advances, a small selection of which are listed below. In general these have involved incremental, rather than quantum jump, improvements and have been achieved by empirical selection and not, as yet, by MAS.

- Many releases of drought and acid soils tolerant tropical maize varieties released worldwide, e.g. the acid soils tolerant CORPOICA H-108 and H-111 for Colombia, the Pool 25 and population 28 lines for acid soils used the Brazilian programme, ZM421, 521 and 621 recently released in Southern Africa which are both tolerant of low nitrogen and mid-season drought. (CIMMYT)
- Rice releases listed as salt tolerant for Bangladesh, e.g. PSBRc 84, 86 and 88. PSBRc 88 has good eating quality and is planted even in non-saline areas
- Drought tolerant banana variety, FHIA01, bred and released in Honduras, now released in Tanzania and in trial in 50 other countries (INIBAP)
- Heat tolerant potato variety, Unica, released in Peru
- Series of durum wheat and barley varieties that have extended the range of these crops in Syria. Chickpea varieties which have facilitated the switch from spring to autumn sown crops (ICARDA)
- Release of drought tolerant Nerica lines, first in Cote d'Ivoire and now particularly in Guinea
- Release of Mulato, a *Brachairia* Al tolerant variety for Mexico and C America.

Progress has however been hampered by the perception that, in some situations, stress tolerance and high yields are incompatible. The view that higher yields under stress conditions are incompatible with higher yields under good conditions invokes the need for independent targeted breeding programmes of specialised varieties. In particular it has been argued that, as drought 'stress intensifies, high yield potential and drought resistance become mutually exclusive'. Counter to this is the conclusion that the improved yield of hybrid maize in the US, where there have been steady improvements since the 1930s, is mainly all the result of selection for response to tolerance to stress and potential yields have not changed over this period. Yet others believe that decentralised participatory breeding with local partners provides the most viable means of breeding locally adapted lines, and at the same time provides an acceptable compromise accommodating the two opposing views. It may well be that there are good physiological reasons as to why the former view may be true for some crops for some stresses at some levels of intensity. Breeding for stress tolerance will proceed more efficiently once it is clear whether, for individual crops and specific stresses, yield potential under stress is controlled by the same genes as yield under optimal conditions. The conclusion will dictate breeding strategy.

APPROPRIATE SECTION SCREENS

Selection screens appropriate to the field conditions that new varieties might experience are often problematic. Field trial climatic factors such as drought and temperature are often unpredictable, while uniform stress conditions are difficult to achieve in trials for edaphic stresses. Also different stresses are often found together, for example salinity problems are rarely all due only to common salt, NaCl. In fact stress the field is rarely due to a single factor. For example multiple metal toxicities and deficiencies are often found simultaneously, and these are farmers' field conditions that are very difficult to match in the laboratory or glasshouse.Of course, in the later stages of a breeding programme, empirical selection under field stress conditions is still probably the available approach for most breeders working in developing countries. The recent developments in thermal imaging and chlorophyll fluorescence imaging may provide rapid, economic, non-invasive selection criteria applications over a range of crops and stress programmes. Good uniform trial sites for stress tolerance selection are not common and where possible should be shared over breeding programmes. The involvement of physiologists with experience of appropriate imaging technologies could benefit a range of CGIAR stress tolerance programmes.

GENOMICS

Developments over the past decade, arising particularly from the human genome programme, have led to a new phase of plant genetics. 'Plant genomics' is the application of the newly available vast amounts of genomic DNA sequence, using a range of novel high-throughput, parallel and other technologies. In plants a 'whole genome' DNA sequence is available as yet only for arabidopsis, which was 'finished' in 2000. A 'draft' raw almost complete sequence of indica rice has been deposited in the public databases by the Beijing group and a similarly complete sequence of Japonica is available within a private company. The fully annotated public DNA sequence of rice, 88% complete at the moment, will be finished later this year. Undoubtedly more species will follow. Possibly maize will be the next major crop plant to be sequenced, at least for 'gene-rich' regions of the genome. Technologies which are included under the umbrella of 'genomics' are: automatic DNA sequencing, where one machine can read two million base-pair a day; microarrays and DNA chips where tens of thousands of genes can be scanned for activity levels at the same time; automated genotyping machines that can assay tens of thousands of DNA diagnostic points a day. In fact it will soon be possible to monitor whole genomes for genetic markers or gene expression on single chips. Transformation technologies that allow the facile and efficient genetic modification of almost all crop plants can also be considered genomics technologies.

Genomics is still in its infancy. Genomics technologies, beyond the now conventional molecular biology technologies, are being taken up by CGIAR Centers and by NARS. High throughput capillary DNA sequencing machines and micro-arrayers are in place in the Centers, transformation as a research tool is available for most mandated species.The CGIAR Task Force for Genomics met in April 2002 to consider the Systemwide accountability and organisation needed for flexible, efficient, sustainable, cost effective genomics for the mandate crops. Some consensus was achieved and more can follow.

SYNTENY AND COMPARATIVE GENOMICS

A second development, which has also emerged over the last decade, is the discovery that gene content and gene order is much more conserved over even quite distantly related species that was previously envisaged. This is known as 'synteny'. Synteny over the grasses, where the phenomenon is most clearly documented, has recently been tested in shotgun rice genomic sequence produced by a private company. Almost all the cereal genes, of known and unknown function, and all of the proteins they code for are found to have corresponding genes in the rice sequence. The maps of the genomes of the grasses, which include all the cereal crop species that have evolved over the past 60 million years, can be aligned so that the location of a gene known in any one gene can be predicted in all the others. A corollary of synteny is, of course, that gene function and role in control of agronomic traits can also be predicted across all the cereals. We should note that the similarity between genomes is restricted to the genes themselves and that intergenic regions differ greatly between even quite closely related species.

The key issue here for the application of genomics tools in the CGIAR crops is that all of the rice resources can be applied directly to the genetic analysis of wheat, maize, barley, pearl millet, finger millet and sorghum. First generation comparative maps have been published for rice and all these genomes. It turns out that the 240 million years that separate the grasses from the broad leafed plants, the eudicots, has degraded precise map correspondence to the point where the retained synteny does not have predictive utility. However the arabidopsis sequence and the arabidopsis genomic resources are available and applicable to broad-leafed crops, e.g. tomato. The genome relationships are close within the Crucifereae, which includes arabidopsis and the brassica crops. The Solanaceae and the Crucifereae are more distant from one another at an estimated 150 million years. Nevertheless the arabidopsis gene organisation can still be used to aid genetic analysis in tomato. The syntenic relationships between arabidopsis and the majority of the broad-leafed CGIAR crops, particularly the legumes, are not well established.However other models are emerging which will aid genetic analyses in these crops. The DNA sequence of a legume model, probably *Medicago* (alfalfa), will be available in the foreseeable future. Also work is progressing rapidly with the tomato genome to bring this species up to model status for all Solanaceous crops.

Breeders and geneticists of cereal crops should become familiar with the relationship of the rice genome organisation and that of their own crop. Breeders of broad-leafed crops where the genome relationships with arabidopsis are not known should have access to the arabidopsis sequence, and should be beginning to establish the syntenic relationships. Novel bioinformatics applications will be are necessary for full application of synteny to crop improvement.

Genomics Applications in CGIAR Mandate Crops

In order to employ genomics to address the problems of abiotic stress in mandate crops and to be able to exploit synteny, rather than simply rely on solutions formulated in the models, a basic genomics infrastructure in the crop itself is required. The bare minimum is probably a molecular framework map of the chromosomes, a large DNA insert library, and a facile transformation system capable of delivering relatively large numbers of engineered plants. The map will have markers every 2 or 3 cM and will have a number of anchor loci, RFLPs or ESTs marked with SSRs or SNPs, that will allow comparisons with the appropriate models. The library will probably be a better than 2-times coverage BAC library with insert sizes in excess of 100 kb.

Additional resources that will be of value include a collection of ESTs, a comparative map and some 'knockout populations. The ESTs will often be comparative collections from stressed and non-stressed plant tissues. The comparative maps will align syntenous chromosome regions of the crop with the model. The knockouts will probably be mutation or deletion libraries in which genes have been disabled at random, although T-DNA tagged or transposon-tagged population, as are available or under development in arabidopsis and rice, are also possible when a good transformation system is available.

Extensive public sector genomics resources are available for maize, wheat and rice. Genomics resources for other CGIAR crops, such as pearl millet, sorghum and *Musa*, are being created in collaborations with ARIs. Other crops such as bean, cassava and *Brachairia* are being worked up within the CGIAR System and yet others, such as forage legumes, beans and chickpeas, have yet to be started.

At the very least base genetic maps should be available for all CGIAR crops. This will open up all species to MAS breeding. Further resources may easily be added at a later date. Links with ARIs or direct outsourcing should be explored for EST and BAC library production.

Gene Cloning

There are many ways to clone genes, however if all that is available is a map location of a stress tolerance gene and there is no indication of its precise function, then the process will probably be either to identify possible candidate genes by their map position or to find genes whose expression is associated with the trait. Certainly in order to find out unequivocally what any gene actually does it will generally be necessary to clone it first.

Map-based cloning

When the alignment of the crop and model maps is available candidate genes mapping in the region of the QTL in the crop can be identified directly from the model DNA sequence. With around 30 genes per map unit, as in rice, it will usually be necessary refine the map location of a QTL, which has been reduced to an identifiable major gene by this time, in large segregating populations. BAC contigs, overlapping linear series of large insert clones, can also be made in the crop itself using DNA landmarks gleaned from both the crop molecular map and the model sequence.

Microarrays

Candidate genes will also emerge from microarray analyses. Genes that are induced by stress are ideal for comparative microarray analysis. A typical experiment will be to challenge an array of ESTs with RNA extracted from stressed and unstressed tissues.

Plainly microarray analysis has a role to play and the CGIAR groups should be gaining experience of the technology. However it is equally clear that microarrays alone will provide complex correlative data which will require considerable further refinement using, for example, knockout phenotypes and QTL location-related map data in order to tease out the causal genetic elements. Candidates from the arrays that coincide with the candidates from map-based approach will be of significant interest.

Comparisons of the DNA sequence or the hypothetical protein product sequence with other isolated genes of known function can provide a lot of information. However while some half of the genes revealed by whole plant genome sequencing remain of unknown function, it will often be necessary to attempt to elucidate function by various reverse and forward genetics methods including transformation and knockout analysis, in both the crop and the model. The definitive experiment will usually involve the production of transgenics with an expected phenotype.

Knockout Populations

Various T-DNA or transposon tagged populations are available in rice and arabidopsis. These reverse genetics 'gene machines' allow the identification of lines in which any gene of interest is disabled. These lines can then be investigated to identify a phenotype that may give clues as to the genes function. A recent development, TILLING allows production of targeted knockouts and also the creation of allelic series in any gene. TILLING populations are available for arabidopsis and are under investigation for rice.

Chemical or irradiation mutant populations in the crop itself will allow a forward genetics approach. Stress tolerant lines can be identified by the simple expedient of subjecting populations to drought, high salt, cold temperatures, submergence etc. and selecting vigorous survivors. The challenge then is to link the phenotype with a deleted gene for which the sequences of candidates will be a good starting point.

With the notable exception of rice, CGIAR germplasm curators have not entered the field of knock-out populations in mandate crops. CGIAR Centers and NARS have significant comparative advantage in having the facilities to grow and maintain large populations, and they have the expertise in the crop to recognise and screen for key knockout phenotypes. These populations will be central to functional genomics efforts and, especially for the minor CGIAR crops, their availability should encourage collaborations with ARI researchers working in model species. The opportunity costs of not taking an international lead of this sort should at least be evaluated for all of the mandate crops.

Transformation in Crop and Model

Genetic transformation is now possible for most crop species. However 'possible' is not the same as 'efficient'. CGIAR Centers need to be able to produce at least tens, and ideally hundreds, of low copy insert transgenics for any construct. Good systems exist for many broad-leafed crops and for most cereals. Legumes are probably the most recalcitrant group of crop plants. Transformation is often seen only as means of making transgenic crops. Indeed, the ideological debate surrounding transgenics notwithstanding, it is inconceivable that we will penetrate far into the 21st Century and its looming food shortages without needing to use all the technology that we have available. However for the time being, transformation has another use as the ultimate test of function of any candidate gene, either in the model or the crop.

Initial attempts will usually employ constructs of the beneficial allele of the gene linked to a constitutive promoter, such as CMV35S, in over-expression experiments. Early transgenic trials will also usually involve antisense constructs that will provide information by negating the effects of the gene. Later trials in the crop itself will probably employ specific alleles of the gene in constructs with promoters that target the effects to specific tissues, such as roots or developing seeds, at particular developmental stages. Transgenics in the crop itself will likely be in an already otherwise adapted genetic background, and these may serve as breeders' lines for eventual introduction into the main stream breeding programmes.

Leads from Model Species

Research into the molecular basis of abiotic stress tolerance is being carried out mainly in model species, particularly arabidopsis. Although this area of our science is still in its infancy there are some 200 references and claims in the reputable scientific press. Genetic transformation experiments to improve stress tolerance are beginning to yield some promising results.

A particularly encouraging approach is the use of transcription factors, regulatory elements that control batches of genes, including those which are induced by stress. One such is *CBF1* in arabidopsis, which is the likely regulator of the cold acclimation response. Over-expression of *CBF1* enhances the levels of a swathe of cold-regulated genes to mimic the effect of cold acclimation that provides subsequent resistance to freezing, and provides protection against cold temperature damage. Transcription factors act as 'master switches' and provide one means of rationalising and exploiting the information obtained from gene expression microarray analyses.

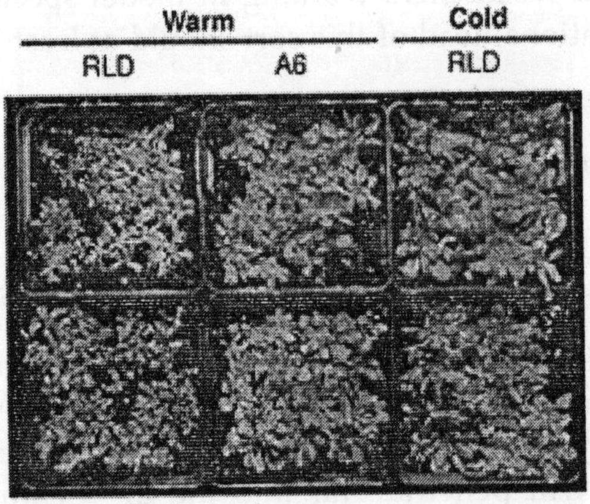

Figure 1. Effect of CBF1 over-expression in arabidopsis, Left: Non- acclimated controls after freezing for 5 days; middle: Non-acclimated transgenics after freezing, right: Acclimated controls after freezing.

DREB1A is another transcription factor that regulates expression of a further range of stress tolerance genes. Over-expression of *DREB1A*, again in arabidopsis, activates expression of a range of genes and results in improved drought, salt and freezing tolerance.

Interestingly it was noted that, when *DREB1A* was driven by CaMV35S, a strong constitutive promoter, normal growth of the plants in an unstressed environment was severely retarded. However the simple expedient of driving *DREB1A* with a stress inducible promoter reduced adverse side effects and further improved tolerance.

Negative pleiotropic effects on fruit yield were also seen in tomato with CaMV35S driven yeast *HAL1* gene, which enhances K^+/Na^+ selectivity and maintenance of water status.

Specific 'ideotype' approaches have also been tried. For example, it has been argued that plants should be able to exploit ions to achieve osmotic adjustment and internally distribute these ions to keep sodium away from the sites of metabolism. To achieve just this a vacuolar Na^+/H^+ antiport, *AtNHX1*, was over-expressed to provide protection up to about half seawater salt levels. Similar effects have been demonstrated with *AtNHX1* over-expression in oil seed rape, *Brassica napus*.

There are many opportunities using the transgenic approach, including, eventually, to produce lines that can be entered into mainstream breeding programmes. Novel genes can be expressed with increasing precision, as more tissue and developmental time specific promoters become available. Endogenous genes can be over-expressed, or negated by the use of antisense constructs.

Interestingly, synteny can also be exploited to good effect. Once the molecular basis of beneficial alleles in any species, has been discovered, including in models like arabidopsis, it is possible to engineer the equivalent homoeologous genes in the target crop with the same alterations in DNA sequence. An excellent example of this approach is the recent targeted engineering of rice to produce a GA-insensitive dwarf phenotype. In 1997 the GAI gene, which produced a gibberellin insensitive dwarf phenotype was isolated from arabidopsis. Soon it was it was possible to demonstrate that rice homoeologues of the arabidopsis gene mapped to locations in cereal genomes that coincided with the location of the 'Green Revolution' wheat semi-dwarfing genes. Moreover the allelic difference between tall and dwarf phenotypes of both rice and wheat were based on the same 51 base-pair deletion in homoeologous genes. Just last year, a Japanese group added the last step when they were able to demonstrate that they could engineer the equivalent rice gene (*SLR*), for which no dwarf mutant had ever been found, with the same 51 bp deletion and produce GA-insensitive dwarf transgenic plants, and thereby avail rice of a completely new and potentially very valuable form of short straw. The same paradigm can, and undoubtedly will, be used to transfer alleles between quite distantly related plants and will no doubt be of value for transferring stress tolerance genes between crops and between wild species and crops.

Finally we should note that the process of transformation is not yet an exact science, as demonstrated by the range of reduced height phenotypes produced by transforming rice with *pSLRtr*. An efficient transformation system will produce a range of 'alleles' from which to choose the 'ideal' phenotype. There appears to a tremendous potential in these results on stress genetics, either in the direct application of model plant gene constructs in crops or in the modification of endogenous crop genes to emulate the effective model plant alleles. The arabidopsis *DREB* genes have already been transferred to wheat with most encouraging effects on salt tolerance. Most CGIAR Centers have effective transformation systems for their own mandate crops and could enter into collaborations to obtain the necessary genes, at least for research purposes.

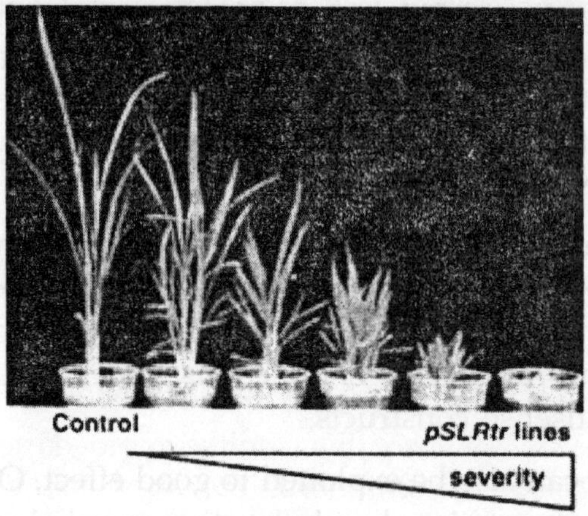

Figure 2. The Same pSLRtr Construct Produces a Range of Transgenic 'alleles'.

Germplasm Collections

The collections held in trust by the CGIAR Centers and by many NARS have major roles to play in both candidate gene discovery and, once genes have been identified as being important in the control of a trait such as stress resistance, identifying the range of alleles available at that locus. It is very clear that these collections will increase in importance. Everything that can be done should be done to provide added value over the next few years.

Allele mining

'Allele mining' will involve PCR-extraction and sequencing of the different versions of genes found in varieties, land races and wild relatives. Variation in gene sequence may then be correlated with the stress tolerance performance of the accession, and may well identify the best alleles for future transgenic experiments.

Association genetics

'Association genetics' is a collection-related development that the CGIAR Centers are particularly well placed to exploit. This new area of science derives from human genetics where analysis of large segregating populations is not possible. Genes associated with any trait are identified by correlation of phenotype with specific alleles at linked markers. In plants this involves scanning collection accessions for variation at marker loci

dispersed over the genome (genotyping) and then correlating, for example, performance under stress of the genotypes with allele dis-equilibrium around the genome. This a young science in plant biology, however the potential is great for discovering novel genes of adaptive significance and for providing added value to the collections. It will not go unnoticed that, once the collection has been 'genotyped', the same data is applicable to any trait of interest.

CGIAR Centers are advantageously positioned to develop plant association genetics. The biodiversity discussion has increased public awareness of the value of *ex situ* collections and any genotyping and further characterisation will increase their value still further. Again the technology and the analyses will be generic so there is much to be gained by close systemwide collaboration.

Abiotic Stress Research in the CGIAR Centers

The arguments developed above indicate that there is a need for a highly co-ordinated pan-crop approach that seeks out and exploits the comparative advantages of all the partners. The approach will also exploit the generic aspects the new genetics and comparative genomics.

The overriding goal must be to produce varieties that will extend the range of arable agriculture and yield more under stressed conditions to provide improved food security for the poor of the world. Sub-goals will include:

- Implementation of an initiative which will underpin stress breeding programmes by providing good science and the best tools to address different problems, without duplication, including:
    - novel genes, and improved versions of genes, to address abiotic stress tolerance
    - improved breeding tools for more efficient incorporation of these genes in new varieties, e.g. better selection screens, better molecular markers, more efficient transformation protocols
    - information on the most appropriate breeding strategies, e.g. whether, for specific crops and specific stresses, breeding can be integral to the core programme, should comprise a separate specialised programme or would be better outsourced to participatory programmes
    - improved knowledge of the physiology and biochemistry underlying stress tolerance.
- Provision of a framework to ensure a continual flow of information between keys ARIs, Centers and NARS. Also the provisions of a forum where ARI scientists can be exposed to the problems encountered in developing countries, and at the same time allow Centers and NARS access to ARI stress science and generic technology.

— Exploitation of the generic experience in molecular genetics and genomics technology, and exploitation of the new opportunities arising from the discovery of synteny between crop genomes, and between crops and models.

— Bringing key skills together to link with breeders to address the problems, e.g. molecular physiology, bioinformatics, genomics technology, IP management etc., particularly where all these skills are not present in a single Center

A successful initiative will exploit the comparative advantages of all potential partners:

— *NAR*—national breeding programmes will have access to the most relevant stress trial sites, increasingly NARS will have molecular biology and genomics expertise, together with skills which are becoming uncommon in the CGIAR, such as physiology and biochemistry. NARS will know local market drivers and will be able to interact will local relevant industry, such as plant breeding and seed companies.

— *ARIs*—Advanced research organisations will provide access to state-of-the - art genomics, and, notably, a few genomics laboratories specialising in abiotic stress. ARI's will provide access to arabidopsis genomics, including the first whole genome arrays, bioinformatics, and biochemical pathway research.

— CGIAR—Centers have a specialised knowledge of the mandate crops, including genetic transformation, availability of field and glasshouse space for trials, and a growing bioinformatics, molecular genetics and genomics capacity. CGIAR Centers hold the key germplasm collections. The Centers have special relationships with NARS and networks in place to transfer technologies

— *Industry*—The multi-national agbiotech industry does undertake some fundamental and strategic and controls some results relevant to Centers' needs. The first whole genome rice arrays are likely to be available from industry. Smaller specialised, often local, companies can provide market-tested genomics service providers, e.g. BAC libraries, DNA sequencing, which will provide benchmarks for Centers planning their in-house/outsourcing research strategy.

The goals and objectives above are embodied in the 'Unlocking genetic diversity in crops for the resource poor' Global Challenge programme. One outstanding recommendation concerning crop replacement should be considered. Although crop improvement to tolerate local will probably be preferred it will be very valuable for agronomists to have available lists of alternative crops that are inherently more tolerant of particular sub-optimal soil types or particular adverse climatic conditions. The compilation of such information for global application would be very valuable indeed.

Genomics and genomics resources in the CGIAR

It is very clear that genomics will play an increasingly important role in CGIAR science and crop improvement. It is equally clear that genomics platform technologies are the most expensive, and probably the most rapidly advancing, that the System has ever had to accommodate. Therefore an early single pan-System policy for the acquisition and deployment of these technologies is imperative.

The goal should be to provide Centers and NARS partner's access to the appropriate genomics resources for all the mandated crops and sustainable access to state-of-the-art platform technologies and plant genomics capacity. An initiative to achieve this would have a number of sub-goals:

- Should be cost-effective and achieve economies of scale while still being flexible and responsive to new developments.
- Providing the CGIAR Centers and NARS access to the rapidly changing state-of-the-art genomics technology.
- Allowing smaller Centers working with wider portfolios of marginal crops to learn from the experiences of the Centers concentrating on the major staples.
- Providing a framework to ensure a continual flow of information between keys ARIs, Centers, NARS and commercial technology suppliers.
- Providing Centers and NARS access to model plant, e.g. rice, *Arabidopsis*, *Medicago*, tomato etc., genomics resources.
- Developing ways of outsourcing *between* Centers
- Providing access to common negotiated sources for standard within-crop genomics services
- Achieve a minimum basic 'in house' genomics infrastructure at the Centers for each crop, e.g. genetic maps, comparative maps with models, BAC libraries, EST collections, efficient transformation methods, genotyped QTL mapping populations (possibly also 'knock-out' libraries), to be shared across the System and with NARS and as vehicles for collaborations with ARIs.

The policy should enable the build-up of pan-System corporate knowledge and capacity of technologies, including:

- ESTs, cDNAs and BAC library production
- marker development - SNPs and, possibly still, SSRs
- high throughput (HTP) genotyping - both for SSRs and SNPs
- association genetics as applied to in-house germplasm collections
- map and comparative map construction and application in mandated crops
- fully genotyped segregating populations for QTL applications

- comparative genomics and bioinformatics, particularly between mandated crops and models
- insertion/mutation populations, Tilling
- handling and storing genomics resources, using laboratory information management systems
- microarrays - both built in-house and 'bought in' Affymetrix-type arrays
- proteomics
- high throughput DNA sequencing
- transformation protocols

All of the above technologies are, or will soon, be required by all Centers. The issue is whether any particular technology is best centralised within Centers, centralised somewhere within the System or outsourced altogether. Centralisation and outsourcing between Centers will certainly become necessary as technology, and associated costs, evolve to soon become beyond the scope of any one Center. Costs are, of course, not the only determining factor. Convenience, service level, relationships between customers and training considerations will all play their part.

Precisely these issues were considered two years ago and reported in the TAC 'Systemwide review of plant breeding methodologies in the CGIAR'. The review emphasized that outsourcing and, especially, outsourcing between Centers should become common for some technologies. The report also noted that centralisation should be considered for technologies that had 'broad utility for all Centers', were 'so expensive that individual Centers cannot afford it', and where' information transfer was synergistic'. In the intervening two years at least two of the CGIAR's genomics technologies have moved into this group.

ESTs, cDNAs and BAC library production should probably all be contracted out, and there are excellent suppliers out there. These are resources that will be revisited over and over, for which quality is paramount. Quality large insert BAC libraries in particular are very reliant on experience and the availability of the appropriate colony picking robots, which will not be present in any Center.

Molecular marker development. This is still a critical activity for many crops. SSRs are generally identified in various enriched libraries or directly from EST sequences. Libraries will probably be made in collaboration with expert ARI labs and the sequencing contracted out directly. SNP detection in the non-staples will again probably be by sequencing PCR copies of cDNA sequences from different varieties. A job for outsourcing. Of course the marker only acquire real value once they have been located in a map framework. This mapping will probably be carried out within Centers.

Genotyping —High throughput genotyping should by now be part of most CGIAR breeding programmes. Using SSRs, although there is trend among industrial groups for global centralisation. These should extend to liquid handling robots for mass PCR and automatic reading of fluorescent labeled products. Proximity to the users - breeders, germplasm curators and geneticists - and the availability of a local system for training purposes argue for in-center systems. These facilities could also be a focus for specific crop NARS breeders' use and training.

All indications are that MAS and germplasm collection characterisation will soon move to HTP SNP genotyping. At present the favoured SNP format has not yet emerged, however when it does it will probably be beyond the financial reach of individual Centers. There are other factors mitigating for centralisation of such a facility. These include large economies of scale and value in having collections of primers at a single site. Also Centers should be incentivised to use a centralised site which represents even a small part of their budget.

A watching brief should be kept on developments, particularly at international breeding companies with interests in multiple crops. Map construction and development of genotyped QTL mapping populations will be carried out within Centers, although the development of crop-model comparative maps will likely be carried out in collaboration with ARIs with expertise in the models.

Mutation and deletion populations, other forms of 'knock-out' lines, possibly 'targeted induced lesions in genomes' (TILLING) populations have not, other than in rice at IRRI, been considered at the Centers. Such genetic stocks play a key role in functional genetics, i.e. assigning function to anonymous gene sequences. The Centers, with their specialised knowledge of the crops and, generally, the space and the manpower to grow large populations under good agronomic conditions, have a comparative advantage in the production of such stocks for the mandated crops. The leveraging power of such resources in the promotion of interactions with ARIs and even companies is obvious. The Centers should carefully consider the opportunity costs of not producing such resources.

Microarrays are of two main types. The first are the high quality, high-density GeneChips produced commercially by a commercial company, Affymetrix, using short 16-32-mer gene-specific oligonucleotides built up on the chip. These include the recently produced 24,000 gene (400,000 spot) whole arabidopsis genome arrays (and, in the near future, barley and rice whole genome arrays, and, in the foreseeable future, wheat, maize etc). The only option at the moment is to buy such chips in (about $700 each or $4,800 including sample processing for a minimal experiment) and process them on an Affymetrix Genechip system costing around $150,000. The second type usually use 'spotted' cDNAs or larger 50-70-mer gene specific oligos and can be made within academic labs. Modern arrayers can work up to 20,000 spots, and cost around $50-100,000.

This sort of facility will usually be associated with significant liquid handling capacity to enable the large numbers of PCR reactions necessary. Also, significant -80°C freezer space will be needed to accommodate the growing amplified cDNA resource. Finally, in a perfect world, one will validate all amplification products by resequencing, so a HTP sequencer may also be required.

It is generally acknowledged that the value of special arrays, such as the rice stress arrays being developed at IRRI, is very dependent on their quality, and this means dedicated expert technical staff associated with a facility. Chip production could be outsourced or centralised within the System, with obvious advantages and disadvantages, but clearly the development of multiple facilities around the Centers is not the best option. Among the advantages the development of quality controlled libraries of cDNAs all in the public sector or for which IP issues are known to have been centrally negotiated. Since it will, for quality control purposes, probably also be necessary to run the hybridisations at the same site, gene expression databases will be developed which are available to all users. A Systemwide facility will be useful for training of other Center and NARS staff. Return on capital outlay is also a factor with an in-house facility. The thinking is that a top-flight arrayer purchased today would remain 'current' for three to four years. Bioinformatics support, both for the direct analysis of results and comparative analyses of the rapidly growing array result databases, is vital and must be factored in to an in-system facility.

Proteomics. Protein analysis, such as performed 'time-of-flight' mass spectrometers is currently outsourced. The cost of the equipment and the rapidly changing state of the art will probably ensure that outsourcing is favoured for the time being.

DNA sequencing. For relatively large-scale sequencing, e.g. several BACs, ESTs in the 1,000s, outsourcing will be the preferred route. This is a very competitive commercial market. Local sequencing capacity within Centers may still be justified for small jobs. Nevertheless, other issues, like the present Indian policy of not allowing DNA to leave the country, may also convince Centers to retain some sequencing capacity.

Laboratory management systems. It will soon become clear that laboratory information management systems (LIMS) are vital for tracking samples and maintain quality control in the laboratory. At present there are none in use around the System (although CIAT are exploring options). The eventual benefits will be large if all Centers were to use the same, or compatible, systems. Genetic transformation is most definitely required in-house and for all crops, although development of the technology may be best carried out in collaboration with ARIs.

The way forward. For cost effective pan-System provision of state-of-the-art genomics infrastructure and technologies iSC might consider the appointment of an independent 'CGIAR Genomics Facilitator'. The initial JD might include:

- a constantly updated trans-national review of outsourced providers and costs for DNA sequencing, DNA library production, proteomics analyses, micro-array facilities and high-throughput genotyping, i.e. constant market testing
- a constant review of outsourcing possibilities between Centers
- act as a clearing house for CGIAR and NARS genomics related queries
- a review of LIMS available, with a view towards harmonisation across the System.
- undertake a special study of the advantages and costs associated with CGIAR centralised micro-array and HTP SNP genotyping services. and, if such facilities move ahead
- the collection of international genomics resources under appropriate MTAs for use with all CGIAR Centers and their stakeholders
- interaction with local managers to establish service level agreements and appropriate financial structures
- commission the development of a web-based tracking system whereby CGIAR and NARS customers can follow the progress of their samples in real time and automatically receive results. This is a key component of any effective and competitive service.

Line management and financial structures will of course need considerable discussion. Operational models like that of Central Advisory Service (CAS) for intellectual property matters at ISNAR should be explored. Also it is possible that the independent facilitator could be closely aligned with, or even be part of, the ISNAR Biotechnology Service (IBS). Plainly a Genomics Facilitator will regularly report and interact on activities through the existing CGIAR Genomics Task Force and the post could report to the System through the chair of that group.

Further activities. Yet another role of a Genomics Facilitator might also assemble and work with multidisciplinary teams, again in actual or virtual centres, across and over crops to address specific issues. The CGIAR Genomics Task-force is an excellent vehicle through which such groupings could meet:

- *Crop type groups.* Cereals, legumes and roots and tuber groups have already been initiated. Certainly the development of stress arrays and anchor markers over related genomes will be crop group activities. These groups will have also specific comparative bioinformatics needs
- *Bioinformatics.* CGIAR over crops bioinformaticists are already linked through Systemwide projects to ARIs in the US and the EU and have skills appropriate to all crops.

International meetings, such as Plant and Animal Genome that is held at San Diego every January, provide excellent opportunities for CGIAR scientists to interact with international academics in:

— *Crop groups.* These already exist and many individual crops are already the subject of international meetings and annual meetings at PAG. A global *Musa* genomics consortium has also recently been formed

Annual meetings organised by CAS at ISNAR or elsewhere for:

— *Intellectual property managers.* Already in place with CAS at the hub, common systems and corporate knowledge, particularly in dealings with industry, are vital. Similarly common IP arrangements should be anticipated across the System for collaborative grants with ARIs, particularly IP for humanitarian use. The Systemwide IP group should also be the preferred partner for NARS.

So finally, rapidly moving research cusps, increasingly expensive technologies, more obvious links between traits and over crops, and increasing technological capacity in NARS are all indicative of more rationalisation, more centralisation, more outsourcing, and more virtual groupings over institutions. The time when we adopt new ways of working cannot be put off much longer.

An appropriate first step forward would be to convene a meeting of the key CGIAR stakeholders to formulate ways in which the new science can be brought to bear in the most efficient manner to deliver the new crops that developing country agricultures need. Such a workshop would be organised by the CGIAR Task Force on Genomics.

Drought, salinity and phosphorus deficiency illustrate the range of abiotic stresses that are faced by farmers in developing countries. Most crop species show considerable genetic variation in tolerance to the major climatic and chemical stresses. Plant breeding is therefore a viable option for improving productivity, reducing farmers' risks and bringing marginal land into use. The CGIAR Centers have a comparative advantage in many aspects of abiotic stress research because of their germplasm collections, their new capacity for genetic and molecular dissection of complex traits, and their ability to conduct multidisciplinary plant improvement programs in target environments. The combined resources of the CGIAR for this work are immense but are underutilised. Investment by the CGIAR in the new tools for gene discovery will produce breakthroughs in our understanding of abiotic stress tolerance that will benefit all the mandated crops. The advances that the Centers are poised to make will find application also in developed countries which are increasingly concerned with the same problems. Improvements in drought tolerance have been responsible for much of the recent increase in yield in maize and have great potential for water saving and risk reduction for rice farmers. Genes responsible for tolerance of salinity, aluminum toxicity and phosphorus deficiency are now close to hand. The impact of these discoveries will be deep and enduring in the communities currently affected by abiotic stress.

Abiotic Stresses for CGIAR Mandate

Abiotic stresses arise from extremes of climate such as drought, flood and cold, from soil toxicities of such elements as Na, Al and Fe, and from soil deficiencies of elements like P and Zn. In places where climatic extremes are of regular occurrence and predictable, agricultural activity is usually very limited and dependent populations are small, but in many other places where these stresses occur in an unpredictable manner the agricultural activity may be intense and the dependent populations large. It is in these latter regions that abiotic stresses are major contributors to food insecurity and poverty for hundreds of millions of the rural poor. Drought alone affects more than 70 million hectares of rice-growing land world-wide. Soil toxicities and deficiencies on the other hand render more than one hundred million hectares of agricultural land marginal for agriculture, again limiting production and creating poverty for millions. Farmers in these environments adopt a risk-aversion strategy of low inputs, resulting in low outputs, poor human nutrition and reduced educational and employment opportunities, especially for girls. The rural poor are particularly badly affected because of lack of access to alternative sources of employment or food.

Plant Breeding for Abiotic Stress Tolerance

Most crop species show considerable genetic variation in tolerance to the major climatic and chemical stresses. Plant breeding is therefore a viable option for improving productivity, reducing farmers' risks and cultivating marginal land. The CGIAR Centers have a comparative advantage in these breeding activities because of their germplasm collections, their new capacity for genetic dissection of complex traits, and their ability to conduct multidisciplinary plant improvement programs in target environments. The Centers have released several high-yielding cultivars with enhanced tolerance of abiotic stresses, including Al-tolerant rice from CIAT, cold-, salt- and submergence-tolerant rice from IRRI, and drought-tolerant maize from CIMMYT. However, these new cultivars represent incremental gains that could certainly be exceeded if more investment were made in the needed multidisciplinary research.

Partnerships for Multidisciplinary Research

It is now a particularly auspicious time for the CGIAR to dedicate itself to a concerted effort to improve tolerance of abiotic stresses in the mandated crops. Advanced research organisations have developed the necessary analytical tools for understanding the mechanisms of stress tolerance. Most of this work is conducted in the public sector, often in basic studies on *Arabidopsis thaliana*. The leading laboratories are eager to be partners in the enterprise. The private sector, with its traditional focus on protecting plants from biotic stresses, at present have little interest in abiotic stresses and may share their new genomic resources with the CGIAR Centers for the benefit of the poor. The linkages between the Centers and their NARES partners are particularly important in defining the research agenda, conducting the breeding programs in realistic environments and ensuring impact.

The early identification of recipient cultivars can greatly accelerate the breeding program and define the baseline performance against which genetic improvement will be judged. The identification of beneficiaries helps in determining the relative contributions of genetic enhancement and crop management to overcoming abiotic stress. Poor farmers will rely more on genetics than management, but expectations should be realistic: it is likely that innovative research on crop and natural resource management in relation to abiotic stresses will reveal cost-effective ways in which poor farmers can increase their productivity and income. Just as the Green Revolution varieties of wheat and rice encouraged massive government infrastructure schemes for irrigated environments, we can expect that new germplasm for the fragile rainfed environments will encourage local innovation in management by and for poor farmers.

Understanding the Mechanisms of Abiotic Stress Tolerance

The screening of germplasm collections for different mechanisms of tolerance to a particular stress must be based on sound physiological principles and take into account the target environment and the timing of stress relative to the growth cycle. Yield under stress is often used as a preliminary criterion that can be applied to thousands of accessions, with more discriminating tests being applied subsequently to identify accessions with different mechanisms of tolerance. If the CGIAR Centers do not invest adequately in the characterisation of the germplasm that they hold in trust, it is unlikely that anyone else will do so and a valuable resource will remain unexploited.

Germplasm accessions with high tolerance of a particular abiotic stress are usually not directly useful for agriculture. The genes conferring stress tolerance must be introgressed into improved backgrounds, a task often rendered difficult by the genetic complexity of the trait and our poor understanding of it at the molecular level. Another limitation is the difficulty of applying a uniform level of stress over a field. Great skill and considerable expense are involved in exposing a population of a thousand breeding lines to uniform stress from drought, salinity, iron toxicity or zinc deficiency. And lack of control of soil type and texture and general climatic conditions can lead to genotype x environment interactions that confound even the most carefully planned experiments. Finally, the target environment is unlikely to feature a single abiotic stress: submergence, drought, iron toxicity and Zn deficiency may be encountered in a single season at a single location. New cultivars with multiple tolerances of abiotic stress are essential.

CGIAR Centers have responded to these challenges by developing interdisciplinary teams focused on specific, high-priority stresses, such as drought, salinity and aluminum toxicity. The powerful new tools of biotechnology have been allied with skills in physiology to design informative experiments. Some blind alleys have been entered, but progress overall has been encouraging. Breakthroughs in one crop are frequently relevant to other crops because of the common background of plant development and metabolism. Studies on one stress may help studies on other stresses because of common principles operating in diverse stress-response pathways.

A major theme in stress biology is the relationship between the evolutionary history of a crop and its whole-plant response to stress. Wild relatives of crop plants generally adopt a fail-safe strategy that, in times of stress, allocates scarce resources to just a few seeds to ensure their vigor as seedlings in the next generation rather than attempting to fill all seeds. In adapting a crop to productive agriculture, it may be necessary to supply just enough crop management or genetic improvement to prevent plants from responding to stress by unnecessarily adopting the same fail-safe strategy. Common principles that apply over species and over stresses are emerging from current studies, raising the possibility of multiple payoffs within the CGIAR system from investments made in any particular crop or stress.

Molecular Tools

Completion of the Arabidopsis and rice genome sequences was announced in 2000 and 2001, respectively. The rate of development of new molecular tools will increase dramatically. For example:

- Microsatellite-based simple sequence repeats (SSR), widely used for marker-aided selection, were until recently quite laborious to find and map. Now they will be available in vast numbers, and few genes of interest will be far from a marker.
- IRRI's activities in proteomic analysis of drought and salt responsiveness in rice led frequently to the detection and partial sequencing of interesting proteins not present in any sequence database. Now the genes corresponding to these proteins reside in databases and wait to be mined.
- Until recently we were unaware of the full complement of plant genes and were ignorant of the genes responsive to any given stress. Now Rice GeneChips carry 24,000 genes and can be regarded as a "mother array" from which trait- or stress-specific "baby arrays" can be readily derived. The "baby arrays" will be cheaper to make and use and easier to interpret.

Other useful molecular tools are also becoming available, such as insertional mutants in rice and maize, and deletional mutants in rice. These resources will allow a direct connection to be made between a gene and a phenotype. IRRI is developing a public platform in functional genomics to which many different institutes will contribute (IRRI itsef, its NARES partners, other CGIAR centers, AROs from developed and developing countries, and the private sector. The position of rice as the model cereal means that this public platform will be useful for the functional genomics activities of CIAT, CIMMYT, ICARDA, ICRISAT and WARDA as well as IRRI.

GENETIC ENGINEERING FOR ABIOTIC STRESS

A number of abnormal environment parameters such as drought, salinity, cold, freezing, high temperature, anoxia, high light intensity and nutrient imbalances etc. are collectively termed as abiotic stresses.Abiotic stresses lead to dehydration or osmotic stress through reduced availability of water for vital cellular functions and maintenance of turgor pressure. Stomata closure, reduced supply of CO_2 and slower rate of biochemical reactions during prolonged periods of dehydration, high light intensity, high and low temperatures lead to high production of Reactive Oxygen Intermediates (ROI) in the chloroplasts causing irreversible cellular damage and photo inhibition.In response to dehydration or osmotic stress a series of compatible osmolytes are accumulated for osmotic adjustment, water retention and free radical scavenging. Similarly, overexpression of certain enzymes such as superoxide dismutase, ascorbate peroxidase and glutathione reductase has been implicated in free radical detoxification and scavenging of free radicals under oxidative stress.

Complexed Stresses by Osmoticum, Dehydration and Salinity

Proline has been recognised as a potent and compatible osmoprotectant which is accumulated in high concentrations in glycophytes and halophytes in response to osmotic stress such as drought and high salinity. Two important enzymes for the biosynthesis of proline i.e. Δ^1 - pyrroline-5-carboxylate synthetase (P5CS) and Δ^1 - pyrroline-5-carboxylate reductase (P5CR) have been cloned from several plants and their expression studied under various abiotic stresses and ABA application.

Some transgenic plants expressing a high level of P5CS mRNA also accumulated high level of P5CS protein. The transgenic plants produced 10 to 18-fold more proline than the control plants. Under drought stress the proline content increased from about 80mg/g fresh leaf to about 3000mg/g in control and from 1000mg/g to an average of 6500mg/g in transgenic lines. Wilting in the transgenic plants was less severe and delayed by 2-3 days in transgenic plants as compared with the wild type (WT) control plants. Their results demonstrated that proline acts as an osmotic protectant and its increased production in the transgenic plants increased tolerance to drought and salt stress.Glycinebetaine, a quaternary amine, is another important compatible solute, which is widely distributed among plants and protects plants on exposure to salt and cold stress. In plants like spinach and barley betaine is synthesized from choline by oxidation of choline to betaine aldehyde and then to betaine. The first step is catalysed by choline mono-oxygenase while the second by nuclear coded gene for betaine aldehyde dehydrogenase. The transformed plants grew slowly at 200 mM NaCl whereas none of the WT plants grew. The accumulation of glycine betaine through genetic engineering in Arabidopsis enhanced its ability to tolerate salt and cold stress.

Anaerobiosis/anoxia

Most plants are highly sensitive to anoxia during submergence. An important aspect of the adaptation to oxygen limitation include metabolic changes such as avoidance of self poisoning and cytoplasmic acidosis, maintenance of adequate supplies of energy and sugar. During anoxia, ATP and NAD^+ are generated not in the Krebs cycle and the respiratory chain but via glycolysis and fermentation. A number of enzymes of the anaerobic pathways such as alcohol dehydrogenase and pyruvate decarboxylase induced during anoxia have been cloned and characterised.

Heavy Metal

Optimum growth and productivity and even cultivation of most of the plants is severely restricted in soils with elevated levels of one or more inorganic ions such as sodium in saline soils; Al, and Mn in acidic soils and heavy metals Cu, Zn Pb, Ni, Cd etc. due to mining, industrial affluents and other human activities.In plants with genetic resistance to Al toxicity, the Al exclusion and uptake from root tips have been found to be correlated to their increased capacity to release organic acids such as citric acid which chelates Al^{3+} outside the plasma membrane. Transgenic tobacco and papaya that overexpressed a citrate synthase gene (CSb) from Pseudomonas aeruginosa in their cytoplasm. Tobacco lines expressing CSb had up to 10-fold higher level of citrate in their root tissues and one of the lines released 4-fold citrate extracellularly whereas in papaya there was only 2 to 3-fold increase of citric acid production. Increased production of citric acid was shown to result in Al tolerance in both the species.

Heat and Cold

Temperate and subtropical plants are highly susceptible to high temperature during early tillering, flower initiation, anthesis and grain filling stages leading to substantial reduction in their productivity. In response to high temperature all organisms, including plants, synthesize a set of proteins called as heat shock proteins (HSPs) which have been classified into several families according to their molecular masses. The induction of HSPs at permissive temperatures have been associated with the acquisition of thermotolerance to withstand short periods of an otherwise lethal temperature.The chilling sensitivity of plants is closely correlated with the degree of unsaturation of fatty acids in the phosphatidylglycerol of chloroplast membranes. Plants with a high proportion of cis-unsaturated fatty acids, such as spinach and Arabidopsis, are resistant to chilling, whereas species like squash with only a small proportion are not. The chloroplast enzyme glycerol-3-phosphate acyltransferase seems to be important for determining the level of phosphatidylglycerol fatty acid unsaturation. Thus they demonstrated for the first time that the level of fatty acid unsaturation of phosphatidylglycerol and the degree of chilling sensitivity of tobacco can be manipulated by transformation with cDNAs for glycerol-3-phosphate acyltransferases from squash and Arabidopsis.

Shading

Optimum supply of nutrients and efficient photosynthesis are conducive to biomass production but the allocation of assimilates within the developing plant determines the harvest index and economic yield. In pure stand canopy as well as in mixed cropping, competition for light energy invokes shade avoidance syndrome manifested by rapid growth and extension of stem and petiole at the expense of leaves, storage and reproductive organs thus predisposing plants to lodging, susceptibility to diseases and insect pests and a lower harvest index. Although the development of semi-dwarf varieties of wheat and rice in the 60s has led to their higher harvest index and grain yield by overcoming some of the defects of the tall genotypes yet the competition among plants for light energy continues to operate in canopies under intensive cultivation practices. The photosynthetic pigments in plants absorb the visible radiation (400-700 nm) and reflect and transmit far red (FR) radiation beyond 700 nm. The FR wave band between 700-800 nm predominating in the dense plant stands have been implicated in proximity perception for initiating shade avoidance syndrome. The FR reflection signals are perceived by the photoreceptors called phytochromes which possess distinct photo sensory functions. Phytochrome (phyA) mediating the inhibition of stem growth on etiolated plants in response to FR wave length 710-720 nm is rapidly degraded and down regulated in light grown plants. Transgenic tobacco lines expressing a high level of heterologous oat phyA apoprotein have been produced.

The level of growth inhibition of transgenic plants correlated with the level of phyA production. Under field trials at various planting densities from 20 to 100 cm, the transgenic plants were indistinguishable from the WT plants at the lowest plant density but became progressively shorter as the plant density increased. This phenomenon termed as "proximity conditional dwarfing" led to a 15 to 20% increase in harvest index (expressed as leaf biomass as a proportion of total biomass) in transgenic plants under high plant density thus demonstrating the suppression of shade avoidance response under high level of phyA expression. Further understanding of the molecular basis of interaction of various phytochromes among themselves and with R: FR rations in natural light environment may help to change crop plant architecture to avoid shade stress and obtain maximum production under high plant density, mixed cropping and agroforestry.

UV - B

The high influxes and absorption of UV-B radiation affects terrestrial plants through damage to DNA directly or indirectly through formation of free radicals, membranes by peroxidation of unsaturated fatty acids, photosystemII, phytohormones and even symbiotic relationship of plants with micro-organisms.A number of secondary metabolites such as flavonoids, tannins and lignins are increased at elevated levels of UV-B radiation which screen UV-B and protect the cellular components against the UV-B damage.

Oxidative Stress

A number of abiotic stresses such as extreme temperatures, high light intensity, osmotic stresses, heavy metals and a number of herbicides and toxins lead to over production of reactive oxygen intermediates (ROI) including H_2O_2 causing extensive cellular damage and inhibition of photosynthesis.

Perspectives and Strategies for Improving Tolerance

The work on genetic engineering of tolerance to abiotic stresses began piece meal within a decade of the molecular understanding of pathways induced in response to one or more of the abiotic stresses. In most of the cases the transgenes expressed faithfully but only a limited level of tolerance was provided under stress conditions as compared to the non-transformed wild type plants. In many cases the transgenic plants had morphological abnormalities and slower growth under nonstressed environment. The level of many compatible osmolytes responsible for osmotic adjustment was too low to be effective per se in providing the required water retention and osmotic adjustment.

Table 1. Genetic engineering of plants for tolerance to abiotic stresses

StressGene/Enzyme	*Source*	
Osmotic	Delta-pyrroling-5-carboxylate synthetase (P5CS)	Mothbean (*V. aconitifolia)*
Drought and Salinity	Mannitol-1-phosphate dehydrogenase (*mt1D*)	*E. coli*
Cold and Salt	Choline oxidase (*cod* A)	*Arthrobactor globiformis*
Salt	Choline dehydrogenase (*bet* A)	*E. coli*
Cold	Omega-3-fatty acid desaturase (*fad* 7)	*Arabidopsis*
Drought	Trehalose-6-phosphate synthase	Yeast
Drought	Levan sucrase (Sac B)	*Bacillus subtilis*

The use of multiple tolerance mechanisms for one or more of the abiotic stresses through stepwise or co-transformation may help to achieve high levels of tolerance for commercial exploitation. The QTL mapping of stress tolerance in certain species, comparative mapping and map based cloning in plants may be used to screen genes which function under stress as well as those induced and expressed in response to stress.

Table 2. Genetic engineering of plants for tolerance to heavy metal

Stress	*Gene/Enzyme*	*Source*
Cadmium	Metallothionein-I (MT-I)	Mouse
Copper	Metallothionein-like (PsMTA)	Pea
Aluminium	Citrate synthase (CSb)	P. *aeruginosa*

Molecular understanding of the stress perception, signal transduction and transcriptional regulation of abiotic stress responsive genes may help to engineer tolerance for multiple stresses. Understanding the molecular mechanism for providing protection against biotic and abiotic stresses may lead to a generalised master mechanism for stress tolerance. Optimum homeostasis is always a key to living organisms for adjusted environments. Thus, abiotic stress accompanying a number of biological phenomena must be precisely investigated by consideration of plant homeostasis.

REFERENCES

Galperin, M.Y., "The molecular biology database collection", *Nucleic Acids Res.* 32, Database issue, 2004.

Graham, C.A. and Hill, A.J.M., (Eds.), *DNA Sequencing Protocols* (2nd ed.), Humana Press, 2001.

Higgins,D. and Taylor,W. (Eds.), *Bioinformatics. Sequence, Structure and Databanks,* Oxford University Press, 2000.

Eggeling, L. and Bott, M., *Corynebacterium glutamicum: Basic Biology and Application, A Handbook,* Boca Raton, FL: CRC Press, 2004.

Nicholl,D.S.T., *An Introduction to Genetic Engineering,* 2nd ed., Cambridge University Press, 2003.

Voet D, Voet J.G., *Biochemistry,* Vol 1 3rd ed. Wiley: Hoboken, NJ, 2004.

Bibliography

Acerenza, L. & Kacser, H., "Enzyme kinetics and metabolic control. A method to test and quantify the effect of enzymic properties on metabolic variables", *Biochem. J.*, 1990.

Barlow, W., "Probable Nature of the Internal Symmetry of Crystals", *Nature*, 1883.

Bendall, D. S., *Evolution From Molecules to Man*, Cambridge, Cambridge University Press, 1983.

Bloomfield,V.A., Crothers,D.M., and Tinoco, I., Jr., *Nucleic Acids. Structures, Properties, and Functions*, University Science Books, 2000.

Bongaerts, J., Krämer, M., Müller, U., Raeven, 1.. and Wubbolts, M., "Metabolic engineering for microbial production of aromatic amino acids and derived compounds", *Metabolic* Engineering, 2001.

Boyer, Rodney, *Concepts in Biochemistry*, Pacific Grove, CA: Brooks/Cole Publishing Company, 1999.

Cai W, Shao X, Maigret B., "Protein-ligand recognition using spherical harmonic molecular surfaces: towards a fast and efficient filter for large virtual throughput screening", *J. Mol. Graph. Model.*, January 2002.

Cowen, Richard, *History of Life*, Boston: Blackwell Scientific, 2004.

Crow, James F. & Motoo Kimura, *Introduction to Population Genetics Theory*, Edina, Minn.: Burgess Publishing Company, 1970.

Dickerson, R.E.; Geis, I., *The structure and action of proteins*, Menlo Park, CA: W.A. Benjamin, 1969.

Dobson CM., "The nature and significance of protein folding", In *Mechanisms of Protein Folding* 2nd ed. Ed. RH Pain, *Frontiers in Molecular Biology series*, Oxford University Press: New York, NY, 2000.

Drexler, K. E., "Molecular Engineering: an approach to the development of general capabilities for molecular manipulation," *Proc. Nat. Acad. Sci.* (USA), Vol 78, 1981, pp. 5275–5278

Eggeling, L. and Bott, M., *Corynebacterium glutamicum: Basic Biology and Application, A Handbook*, Boca Raton, FL: CRC Press, 2004.

Eldredge, Niles, *Macroevolutionary Dynamics*, New York: McGraw-Hill, 1989.

Feynman, R., "There's Plenty of Room at the Bottom," in: Gilbert, H. D. (ed.), *Miniaturization*, Reinhold, New York, 1961, pp. 282–296

Fitch, W. M., "Molecular Evolutionary Clocks, *in* Ayala, F. J., ed., *Molecular Evolution: Sunderland*, Mass., Sinauer, 1976.

Flint, H. J., Dible, S. & Kacser, H., "Derepression of enzyme synthesis is response to arginine limitation in *Neurospora crassa*", *J. Gen. Microbiol.*, 1985.

Fruton, Joseph. *Proteins, Genes, Enzymes: The Interplay of Chemistry and Biology*. New Haven: Yale University Press. 1999. ISBN 0-300-07608-8

Futuyma, Douglas J., *Evolutionary Biology*, Sunderland, Mass.: Sinauer Associates, 1997.

Galperin, M.Y., "The molecular biology database collection", *Nucleic Acids Res.* 32, Database issue, 2004.

Gillespie, John H., *The Causes of Molecular Evolution*, New York: Oxford Univ. Press, 1997.

Glover, D. M., and Hames, B. D., *Genes and Embryos*, New York, Oxford University Press, 1989.

Golding, Brian, ed., *Non-Neutral Evolution*, Boston: Chapman and Hall, 1994.

Graham, C.A. and Hill, A.J.M., (Eds.), *DNA Sequencing Protocols* (2nd ed.), Humana Press, 2001.

Graur, Dan & Wen-Hsiung Li, *Fundamentals of Molecular Evolution*, Sunderland, Mass.: Sinauer Associates, 2000.

Hartl, Daniel L. Andrew G. Clark, *Principles of Population Genetics*, Sunderland, Mass.: Sinauer Associates, 1997.

Higgins,D. and Taylor,W. (Eds.), *Bioinformatics. Sequence, Structure and Databanks*, Oxford University Press, 2000.

Ikeda, M., "Amino acid production processes", *Advances Biochernical Engineering*, 2003.

Jorde, L.B., J.C. Carey, M.J. Bamshad, and R.L. White, *Medical Genetics*, 2nd ed. St. Louis: Mosby-Year Book, Inc., 2000.

Kacser, H., "The control of enzyme systems *in vivo*", *Elasticity analysis of the steady state, Biochem. Soc. Trans.*, 1983.

Kimura, Motoo, *The Neutral Theory of Molecular Evolution*, Cambridge, U.K.: Cambridge Univ. Press, 1983.

Kinloch A.J., *Adhesion and Adhesives: Science and Technology*, Chapman and Hall, 1987.

Kuhlman B, Dantas G, Ireton G.C, Varani G, Stoddard B.L, Baker D., "Design of a novel globular protein fold with atomic-level accuracy", *Science*, 2003.

Lewontin, Richard C., *The Genetic Basis of Evolutionary Change*, New York: Columbia Univ. Press, 1974.

Lily E. Kay, *The Molecular Vision of Life: Caltech, the Rockefeller Foundation, and the Rise of the New Biology*, Oxford University Press, Reprint 1996

Lodish, H., et. al., *Molecular Cell Biology*, 4th ed. New York: W. H. Freeman & Co., 2000.

Middleton, R. J. & Kacser, H., "Enzyme variation, metabolic flux and fitness: alcohol dehydrogenase in *D. melanogaster*", *Genetics*, 1983.

Morange, Michel. *A History of Molecular Biology*. Cambridge, MA: Harvard University Press. 1998.

Morris R.J, Najmanovich R.J, Kahraman A, Thornton J.M., "Real spherical harmonic expansion coefficients as 3D shape descriptors for protein binding pocket and ligand comparisons", *Bioinformatics*, May 2005.

Nelson, David L., David L. Nelson, and Michael M. Cox, *Lehninger Principles of Biochemistry*, 3rd edition. Worth Publishing, 2000.

Nicholl,D.S.T., *An Introduction to Genetic Engineering*, 2nd ed., Cambridge University Press, 2003.

Noel S. Hush, "An Overview of the First Half-Century of Molecular Electronics", *Ann. N.Y. Acad. Sci.*, 2003.

Ohnishi, J., Mitsuhashi, S., Llayashi, M., et al., "A novel methodology employing *Corynebacterium glutamicum* genome information to generate a new r-lysineproducing mutant", *Appliecl Microbiology and Biotechnology*, 2002.

Provine, William B., *The Origins of Theoretical Population Genetics*, Chicago: Univ. of Chicago Press, 2001.

Rastetter, W. H. "Enzyme Engineering," *Applied Biochem. and Biotech.*, Vol. 8, 1983, pp.423–436

Sauro, H. M. & Kacser, H., "Enzyme-enzyme interactions and control analysis", 2. The case of non-independence: heterologous associations, *Eur. J. Biochem.*, 1990.

Thieffry, D., "Forty years under the central dogma", *Trends Biochem. Sci.*, 1998.

Voet D, Voet J.G., *Biochemistry*, Vol 1 3rd ed. Wiley: Hoboken, NJ, 2004.

Walian P, Cross T.A, Jap B.K., "Structural genomics of membrane proteins", *Genome Biology*, 2004.

Watson, J.D. and Crick, F.H.C., "Molecular structure of nucleic acids", *Nature*, 1953.

Weiner, S. J., et al., "A New Force Field for Molecular Mechanical Simulation of Nucleic Acids and Proteins," *J. Am. Chem. Soc.*, Vol. 106, 1984, pp. 765–784.

Index